7th International Conference on Applications of Physics in Financial Analysis 2009

(APFA7)

Journal of Physics: Conference Series Volume 221

Tokyo, Japan
1-5 March 2009

Editors:

Misako Takayasu
Yuichi Ikeda
Tsutomu Watanabe
Hideki Takayasu

ISBN: 978-1-61738-655-8
ISSN: 1742-6588

Printed from e-media with permission by:

Curran Associates, Inc.
57 Morehouse Lane
Red Hook, NY 12571

Printed by Curran Associates, Inc. (2010)

Institute of Physics
Dirac House, Temple Back
Bristol BS1 6BE UK

Phone: 44 1 17 929 7481
Fax: 44 1 17 920 0979

techtracking@iop.org

Additional copies of this publication are available from:

Curran Associates, Inc.
57 Morehouse Lane
Red Hook, NY 12571 USA
Phone: 845-758-0400
Fax: 845-758-2634
Email: curran@proceedings.com
Web: www.proceedings.com

TABLE OF CONTENTS

This volume contains contributed papers from the 7th international conference on 'Applications of Physics in Financial Analysis (APFA)' held at Tokyo on 1–5 March 2009. The conference was organized jointly by Tokyo Institute of Technology and Hitotsubashi University with support from the Research Institute of Economy, Trade, and Industry (RIETI), Physical Society of Japan, Japanese Economic Association, Information Processing Society of Japan, Japanese Society for Artificial Intelligence, and Japan Association for Evolutionary Economics. The first APFA conference (APFA1) was held in 1999 at Dublin, followed by APFA2 at Liege in 2000, APFA3 at London in 2001, APFA4 at Warsaw in 2003, APFA5 at Torino in 2006, and APFA6 at Lisbon in 2007. The 7th APFA conference, which is the first meeting held outside Europe, was attended by 223 researchers in physics and economics from 23 countries world-wide. In keeping with past APFA conferences, we paid special attention to issues in financial markets, which turned out to be very timely. The conference was held in March 2009, in the middle of the global financial crisis that originally started in the US and spread quickly to every corner of the world.

The topic of the conference is 'New Approaches to the Analysis of Large Scale Business and Economic data'. The rapid development of information and communication technology has enabled financial/non-financial firms to keep detailed records of their business activities in the form of, for example, tick-by-tick data in financial markets, point-of-sale (POS) data on individual household's purchasing activity, and interfirm network data describing relationships among firms in terms of suppliers/customers transactions and ownerships. This growth in the scope and amount of business data available to researchers has led to a far-reaching expansion in research possibilities. Researchers not only in social sciences but also in physics, mathematics, and information sciences have recently become interested in such datasets, conducting empirical investigations about various aspects of economic activities. Specifically, they have searched for regularities and 'laws' akin to the ones in natural science, successfully producing fascinating results, as shown in the papers contained in this volume.

Each paper submitted for publication in this volume has gone through the refereeing process, and has been revised on the basis of comments and discussion at the conference as well as comments from the anonymous referees. Finally, 19 papers were accepted for publication. The editors are very grateful to the colleagues involved in the refereeing process for their rapid and careful reviewing of the papers.

We thank Takayuki Mizuno, Koji Sakai, Hiwon Yoon and Hiroki Matsui for their support for the conference. We appreciate the administrative assistance provided by Yayoi Hatano of Hitotsubashi University, and Masahiko Ozaki, Masato Yamada and Tomoko Kase of RIETI. We are most grateful to the authors for their contributions, as well as to the participants, all of whom made this conference stimulating and enjoyable.

Misako Takayasu
Tokyo Institute of Technology, Japan

Tsutomu Watanabe
Hitotsubashi University, Japan
RIETI, Japan

Yuichi Ikeda
Hitachi Research Laboratory, Hitachi Ltd, Japan

Hideki Takayasu
Sony Computer Science Laboratories, Inc, Japan

Rank-correlations and value-correlations of Gross Domestic Product per Capita in Latin American Countries

Francisco O. Redelico[1,2], **Araceli N. Proto** [1,3], **Paulette Clippe** [4] **and Marcel Ausloos** [45]

[1] Facultad de Ciencias Fisicomatemáticas e Ingeniería, Universidad Católica Argentina.
[2] Laboratorio de Sistemas Complejos, Facultad de Ingeniería, Universidad de Buenos Aires, Argentina.
[3] Comisión de Investigaciones Científicas, Provincia de Buenos Aires, Argentina.
[4] GRAPES@SUPRATECS, Institute of Physics, B5a, ULG, B-4000 Liège, Euroland

E-mail: francisco_redelico@uca.edu.ar

Abstract. This work describes a method for searching for globalization evidence within Latin American countries using correlation networks methods. Two correlation measures are used, one based on the usual Pearsons Correlation Coefficient and the other based on Mutual Information. First, it is pointed out there is a *core* of globalization, where no trade blocs appear, within Latin American countries and second, a hierarchy, from a globalization point of view, is found within these countries. There is no intention to enter into a political consideration here, though any politically prone reader may guess that some further consideration is in order.

1. Introduction

In previous works [1] the *evolution* of the Gross Domestic Product per Capita (*GDP*) of Latin American countries was analyzed from an expansion and crisis point of view. Moreover the *GDP* evolution for rich countries has been quantified as well [2] using techniques from statistical mechanics like the well known complex network paradigm [3] or correlation entropy-like measures ([4, 5]).

This field of investigations is of interest because modeling coordination of macroeconomics variables has an important role in econophysics. One can hope that this activity can help economists and politicians to better appreciate numbers and perform some better management. Therefore our contribution is devoted to give some insight, methodology and technique, on how aspects of macroeconomic questions could be analyzed using statistical concepts and complex networks, whence *in fine* how to find out whether societies and economies are interrelated in a (nonlinear) way.

As globalization can be described as a process in economy with increasigly similar features in developping patterns, hence the Gross Domestic Product *per capita* will be presented in order to establish a degree of globalization of macroeconomy. There is no intention to enter into a political consideration here, though any politically prone reader may guess that some further

[5] Present address:7, rue des chartreus, B-4122 Plainevaux, Belgium

consideration is in order.
The paper is organized as follows: Section 2 presents the analyzed data set and the Latin America country list hereby examined. Section 3 presents the methodology used in order to perform an analysis of such (GDP) data through a weighted network analysis and Section 4 presents two examples of such results. A brief summary is given in the final section to serve as conclusion.

2. Data

The data set used is the same as in [1, 2], the countries considered for the analysis are those Latin American ones where Spanish and Portuguese languages prevail, whence countries like Belize, Jamaica, Barbados, Trinidad and Tobago, Guyana, Antigua and Barbuda and the Bahamas, where English prevails, Suriname, Netherlands Antilles and Aruba where Papiamento prevails, Haiti, Martinique and Guadeloupe and French Guiana where French prevails are excluded from the present analysis. All GDP data is freely provided by the Development Data Group of The World Bank [6]. As The World Bank [6] does not report any data about Cuba, this country is excluded from the analysis, although Spanish is its official language. Therefore only $M = 19$ countries are considered in the present analysis. This set of data will be referred to as LA19 data. This definition is somewhat arbitrary, but it preserves some homogeneity among the LA country selection and is reported in [7]. According to that, only annual records of GDP/capita for Argentina (ARG), Bolivia (BOL), Brazil (BRA), Chile (CHL), Colombia (COL), Costa Rica (CRI), República Dominicana (DOM), Ecuador (ECU), El Salvador (SLV), Guatemala (GTM), Honduras (HND), México (MEX), Nicaragua (NIV), Panama (PAN), Paraguay (PRY), Peru (PER), Puerto Rico (PRI), Uruguay (URY) and República Bolivariana de Venezuela (VEN) between 1960 and 2006 can be examined; the country name is abbreviated according to The Roots Web Surname List (RSL) [8] which uses 3 letters standardized abbreviations to designate countries and other regional locations. Since in the case of Puerto Rico the data is available only up to 2001, in order to keep some coherence, all temporal series for the other countries are also limited to the first $N = 42$ data points.

3. Methodology

Three methods are applied in order to find synchronization of the *GDP/capita* fluctuation between countries, i.e. (i) the usual Pearson correlation coefficient, (ii) a nonlinear correlation coefficient and (iii) the Average Overlapping Index (AOI). The information correlation entropy is given. Finally endowed with these measures of similarity evolving networks are constructed. According to criteria based on AOI and some so called mean distance threshold we showed the most relevant countries.

3.1. Pearson Correlation Coefficient (PCC)

In order to get a linear measure of correlation for the evolution of the *GDP/capita* for each pair of countries, the well known Pearson's correlation coefficient is used. This coefficient is a parametric statistics depending on the first two moments of the probability distribution governing the behavior of the time series analyzed and calculated using the *values* of each of the time series involved. It is necessary to be aware that most of the macro economical time series present nonstationary characteristics, and some quite finite size. Therefore it is recommended that the correlation patterns should be investigated by moving a constant size time window (TW) with a constant step so that the whole time interval is scanned as in [2]. Thus we use:

$$C_{ij}(t, TW) = \frac{< x_i x_j >_T - < x_i >_T < x_j >_T}{\sqrt{< x_i^2 - < x_i >_T^2 >_T < x_j^2 - < x_j >_T^2 >_T}} \tag{1}$$

where x_i and x_j, $i, j = 1, ..., M$ are the GDP/capita yearly fluctuation data series for two countries i and j; t is the end of the time interval considered $[t - TW; t]$; $\langle \cdot \rangle_T$ means averaging over the time window of size TW so considered.

In order to get a measure of the correlation for the whole time span we need to convert the *PCC* into an additive measure [2], i.e. the square of the C_{ij}'s is taken before averaging, to obtain the so called *coefficients of determination* (C_{ij}^2) which are additive. In doing so, the PCC over the whole time span is:

$$\hat{C}_{i,j}(TW) = \left[\frac{1}{\nu} \sum_{t=k}^{k+TW} C_{i,j}^2(t) \right]^{1/2} \tag{2}$$

where t is a discrete counter variable; $k = 0, 1, \ldots, N - TW$; TW is the time window size used for the analysis; in this definition $\nu = N - TW + 1$, N being the total number of times (points) at which some measure of x_i is available.

The values of the PCC, Eq. (2), are used as the weights $w_{i,j}^{linear}(TW)$ for the links of the network to be constructed with M nodes through an adjacency matrix with elements

$$w_{i,j}^{linear}(TW) \equiv \hat{C}_{i,j}(TW) \tag{3}$$

obviously with $w_{i,i}^{linear} = 0$. The mean (over the network) weights are defined as

$$\left\langle w_{i,j}^{linear}(TW) \right\rangle = \frac{1}{M(M-1)} \sum_{i=1}^{M} \sum_{j=1}^{M} w_{i,j}^{linear}(TW) \tag{4}$$

3.2. Nonlinear Correlation Coefficient (NCC)
To overcome drawbacks generated by the dependence of the PCC to the first and second moments and in order to grasp correlation others than linear [4] a nonlinear correlation coefficient (NCC) is used. This is a distribution-free statistics [9] coefficient, based on *ranks*, depending on a parameter b that has no *a priori* interpretation; it should be considered as a tuning parameter in order to vary the NCC computation, as it is usual in most nonparametric statistics methods.

For multivalue or multivariate data a correlation matrix can be formed using the NCC, - as defined by Eq. (8) of [5],

$$w_{i,j}^{nonlinear}(b) \equiv NCC_{i,j}(b) \tag{5}$$

where $NCC_{i,j}(b)$ is the nonlinear coefficient between the *i-th* and *j-th* variable. In (Eq. 5), b plays the same parametric role as TW in (Eq. 1).

The mean weight is defined by

$$\left\langle w_{i,j}^{nonlinear}(b) \right\rangle = \frac{1}{M(M-1)} \sum_{i}^{M} \sum_{j}^{M} w_{i,j}^{nonlinear}(b). \tag{6}$$

since $w_{i,j}^{nonlinear}(b) = 0$.

3.3. Overlapping Index
Up to now, only measures quantifying similarities for the GDP/capita fluctuation between each pair of countries are explained. A statistics named *Average Overlapping Index* has been proposed to quantify the informational power of each node. Let us briefly recall the AOI mathematical

formulation and its properties in the case of an unweighted network made of M nodes linked by edges (l, m) $(1 \le l \le M; 1 \le m \le M)$. The *overlapping index* is defined as

$$O_{lm} = \frac{N_{lm}\,(k_l + k_m)}{4\,(M-1)\,(M-2)}, \qquad l \neq m \tag{7}$$

where N_{lm} is the measure of the common number of neighbors of l and m nodes[6] and k_l is the *vertex degree* or the number of vertex connections of node l. Eq.(7) can be generalized for a weighted network as:

$$O_{ij} = A \sum_{\substack{l=1 \\ l \neq j,i}}^{M} (w_{il} + w_{jl}) \left(\sum_{\substack{p=1 \\ p \neq i}}^{M} w_{ip} + \sum_{\substack{q=1 \\ q \neq j}}^{M} w_{jq} \right) \tag{8}$$

where

$$A = 1/[4\,(M-1)\,(M-2)] \qquad 0 < O_{lm} \le 1, \tag{9}$$

in which $w_{i,j}$ is either taken from the PCC or NCC data base. The *Average Overlap Index* for the node (country) l is defined as:

$$\langle O_l \rangle = \frac{1}{M-1} \sum_{m=1}^{M} O_{lm}. \tag{10}$$

Of course, $O_{ll} = 0$.

The AOI values for each country as calculated from the linear correlation matrix are shown in Fig. 3 and for the nonlinear correlation matrix, in Fig. 2, in AOI value vs. rank plots.

3.4. Information correlation entropy

The *Information Correlation Entropy* defined as:

$$I = 1 + \sum_{i=1}^{N} \frac{\lambda_i}{N} \log_N \frac{\lambda_i}{N} \tag{11}$$

where λ_i is the *ith* eigenvalue of the given correlation matrix. The behavior of the information entropy for both the linear and the nonlinear correlation matrices are presented in Fig 1 a function of the time window size which can be used.

Fig. 1 seems to indicate that a sort of phase-transition exists for such data analysis, i.e. an optimal $TW = 15$ could be chosen for the NCC and in this TW the entropy of the PCC gets its local maximum. This phase-transition-like result in the evolution of the GDP/capita for Latin American countries agrees with that obtained in [1] using critical exponent theory.

4. Networks

Endowed with the measures explained above, the procedure described in [4] can be used to construct the correlation matrix between the GDP time series. In order to construct their respective adjacency matrices a method similar as proposed in [4] can be applied. This method consists in assigning to each node its respective AOI, as an *informational measure*; the link between each pair of nodes corresponding to the weight taken from the correlation matrix.

[6] in a fully connected network, $N_{lm} = M - 2$

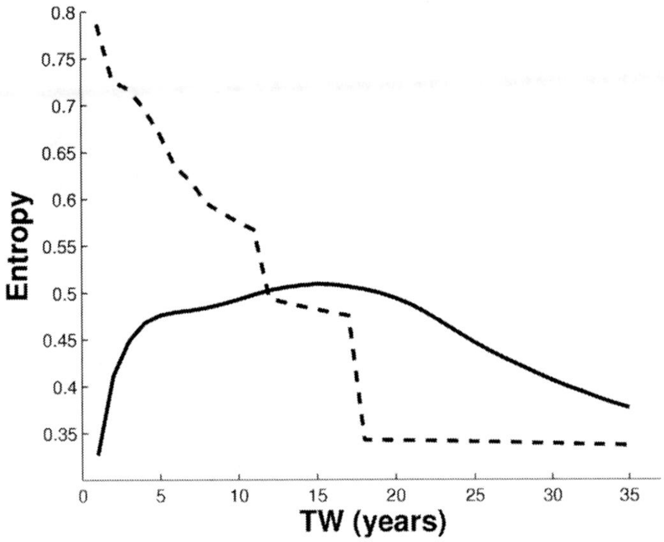

Figure 1. The correlation information entropy I_{R^N} (Eq. 11) for Latin American countries correlation matrix vs. the time windows (TW) size. Dashed-line: correlation information entropy for LA countries from the linear correlation matrix, solid-line: correlation information entropy for LA countries from the nonlinear correlation matrix.

First, a threshold can be imposed on the AOI scale in order to maintain the highest AOI nodes in the network structure. Next, only links higher than a threshold can be kept in the network topology. This allows combining node [10] and link [11] attrition methods for finding network evolution and network structure evolutions, as a function of their internal statistical characteristics. Of course, these thresholds should be carefully selected. If the former threshold is too high only a few nodes remain in the network and if the latter is too low, links which in fact contain "*noise*" or irrelevant information may still be present.

Two examples are shown to illustrate these points.

- A network generated from the PCC matrix only with the countries which have an AOI greater than the third quarter of their $w_{i,j}^{linear}$ set, i.e. AOI $\geq$0.43, and having a link threshold greater than $\langle w_{i,j}\,(TW = 15)\rangle = 0.73$ is presented in Fig. 5.
- A network generated from the NCC matrix only with the countries which have an AOI greater than the third quarter of their $w_{i,j}^{nonlinear}$ set, i.e. AOI ≥ 0.45, and the threshold used for removing weak links was $\langle w_{i,j}\,(b = 15)\rangle = 0.66$, as defined in Eq.(6) is presented in Fig. 4.

5. Conclusions

This contribution is a starting point to enhance some feature in a macroeconomics context. We can state that we do better than classical numerical models like used in classical time series analysis. Our argument for example pertains to the sort of network left over after the node-link attrition process. One notices that COL, PRI, DOM and CRI are the countries having the highest AOI in the PCC and NCC approach, but do not necessarily form a complete graph, - see the missing link between DOM and CRI in the case of NCC and between GTM, PRI and DOM in the PCC case. Intuitively one can imagine some economic role for COL, and observe the geographical position of the three other countries as well. An open question is the role of

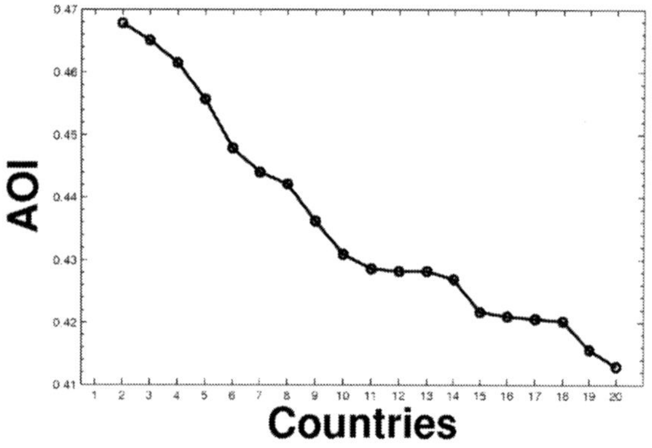

Figure 2. This plot shows the order of countries according its AOI when the correlation matrix is calculated using the nonlinear correlation coefficient (NCC). Selected countries in order to construct the network are the first 7 major AOI.X axis reads 1: COL, 2: PRI, 3: MEX, 4: DOM, 5: CRI, 6: PRY, 7: PAN, 8: SLV, 9: GTM, 10: ECU, 11: HND, 12: BRA, 13: PAN, 14: BOL, 15: ARG, 16: NIC, 17: CHL, 18: PER, 19: VEN.

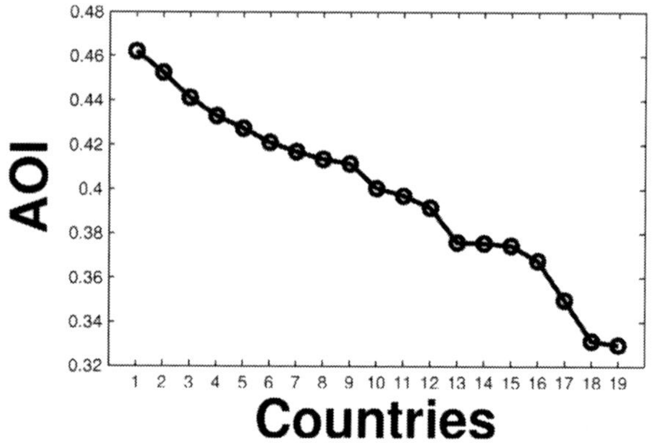

Figure 3. This plot shows the order of countries according its AOI when the correlation matrix is calculated using the Pearsons correlation coefficient (PCC). Selected countries in order to construct the network are the first 5 major AOI.X axis reads 1: COL, 2:DOM, 3: PRI, 4: GTM, 5: CRI, 6: PRY, 7: PAN, 8: BRA, 9: MEX, 10: HND, 11: CHL, 12: NIC, 13: ECU, 14: SLV, 15: URY, 16: BOL, 17: VEN,18: ARG,19: PER.

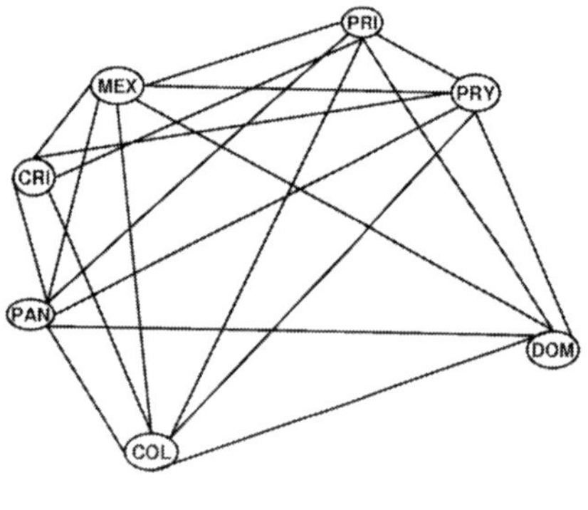

Figure 4. Network generated from the NCC matrix only with the countries which have an AOI greater than the third quarter of their $w_{i,j}^{nonlinear}$ set, i.e. AOI $\geq$ 0.45, the threshold used for the links was $\langle w_{i,j}\,(b = 15)\rangle$ as defined in Eq.(6); recall that its value is 0.62. The lengths of the links between nodes have no other meanings than indicating connections.

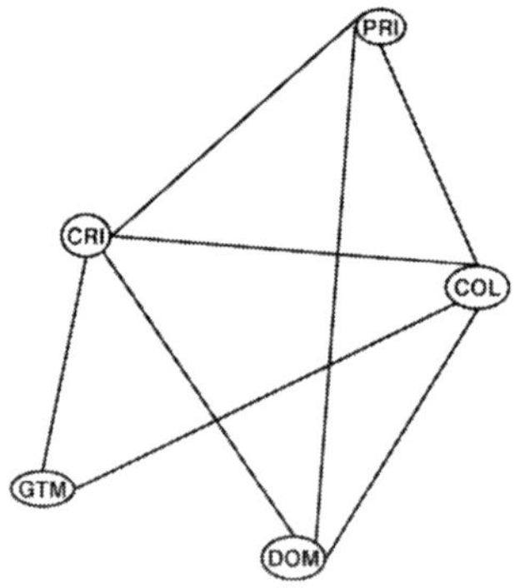

Figure 5. Network generated from the PCC matrix only with the countries which have an AOI greater than the third quarter of their $w_{i,j}^{linear}$ set, i.e. AOI $\geq$0.43, the threshold used for the links was $\langle w_{i,j}\,(TW = 15)\rangle$ as defined in Eq.(4); recall that its value is 0.63. The lengths of the links between nodes have no other meanings than indicating connections.

PRY seen through the NCC approach, that of MEX, GTM, and PAN being more geopolitically understandable.

The most important conclusions in this contribution are: first, none of the trade blocs formed by

Latin American countries belong to the final networks Fig. 4 and second, looking for a hierarchy within Latin American countries GDP/evolution, the highest AOI is Colombia, it might reflects it relationship with United States of America and the lowest AOI is for Venezuela, might due it is the largest petroleum exporter in Latin American.
Thus it seems that our approach indicates many other lines for work, -like systematic investigations on perturbations, i.e. like on the role of non-latin american countries in the evolution and structures of such LA network features, etc. Our statistical approach suggests also ways to develop models based on clusters as in condensed matter Hamiltonian physics, etc.

Acknowledgments
Thanks to a CONICET-CGRI -FNRS bilateral agreement which has allowed interaction between the authors. FOR gratefully thanks APFA7 Organizer Comitee for substantial financial support to attend the meeting and Universidad Católica Argentina for partial financial support. The present work was also supported by BID OC-AR 1728, PICT 02 13533.

References

[1] F.O.Redelico, A.N.Proto and M.Ausloos, Physica A **387** (2008) 6330-6336.
[2] M. Gligor and M. Ausloos, Eur. Phys. J. B **63** (2008) 533-539
[3] R. Albert and A.-L. Barabasi, Rev. Mod. Phys. **74** (2002) 47-96
[4] F.O.Redelico and A.N.Proto, Int. J. Bifurcation Chaos (in press)
[5] Q. Wang, Y. Shen. and J.Q. Zhang, Physica D **200** (2005) 287-295.
[6] *http://go.worldbank.org*
[7] The New Oxford Dictionary of English. Pearsall, J., ed. 2001. Oxford, UK: Oxford University Press; p. 1040
[8] *http://helpdesk.rootsweb.com/codes/.*
[9] J.S. Maritz, Distribution-Free Statistical Method. 2nd Edition (Monograph on Statistics and Applied Probability 17) Chapman & Hall. UK, 1995.
[10] I. Hellsten, R. Lambiotte, A. Scharnhorst, M. Ausloos, Scientometrics **72** (2007) 469-486.
[11] R. Lambiotte, M. Ausloos, Eur. J. Phys. B **50** (2006) 183-188.

Fractal geometry of prior-to-crash market situations: The NASDAQ100 April 2000 crash case.

Francisco O. Redelico[1,2] **and María N. Piacquadio**[3]

[1] Facultad de Ciencias Fisicomatemáticas e Ingeniería, Universidad Católica Argentina.
[2] Laboratorio de Sistemas Complejos, Facultad de Ingeniería, Universidad de Buenos Aires, Argentina.
[3]Secretaría de Investigación y Doctorado, Facultad de Ingeniería, Universidad de Buenos Aires,Argentina.

E-mail: `francisco_redelico@uca.edu.ar`

Abstract. The search for laws governing the occurrence of financial crashes is one of the most important activities in the field of econophysics. In this contribution the daily fluctuation of the price at the close time series before the April 2000 crash NASDAQ100 is analyzed using fractal geometry, two multifractal spectra are derived and a new instability index is presented in order to enlighten the results obtained within these spectra. Using these methods, empirical evidence of the usefullness of the discrete Chakrabarti-Stinchcombre model for earthquakes to describing the rising part of a speculative buble leading a crash is found.

1. Introduction

The pioneer work of Benoit Mandelbrot about fractals was about studying income distribution in a society [1]. He found the fluctuations at global scale (cotton market) are also presented at small scale (individual incomes) [2]. This is the simplest notion on which relays the fractal nature of most of the real problems that appear in economic as well as biological systems. The most popular way to describe geometrical objects within the fractal geometry notion is the box counting dimension. In previous papers [3],[4] it was pointed out that there is many huge fluctuation in the prices of the stock market in a pre-crash situation. The aim of this paper is to characterize the fractal (or multifractal) structure of these fluctuations, based in a fluctuation time series (i.e. $log(price\,(t+1) - price\,(t))$). It is well known that financial and economic phenomena can be enclosed in this geometry [5], so this geometry will be used in this paper in order to describe the prior to crash evolution of the daily fluctuation of the price at the close series from NASDAQ 100 composite index April 2000 and a period far from any large crash (i.e. DJIA 1990-1994).

Several researchs [6, 7, 8] have pointed out formal similarities between eathquakes magnitudes time series and the stock market price evolution, leading to the Chakrabarti-Stinchcombre model [6]. The well known NASDAQ100 composite index April 2000 crash ([4] and he references therein) is used in order to show the appearance of the two overlapped Cantor set in the rise branch of the speculative bubble as is modelled in the aforementioned model.

This paper reads as follow: Section 2 is devoted to explain both multifractal spectra calculated from the time series analyzed, Section 3 presents the new instability index and finally

Section 4 presents conclusions.

2. Data and methodology

As it was told in Section 1 the box-counting dimension is the most popular measure of the dimension in fractal geometry. Unfortunately it does not give good enough quantitative description about a measure supported by a set with multifractal characteristics (i.e. different dimension in every scale). In order to take the problem of an infinitely number of dimensions, the measure is normalized to be a probability distribution. The support of this measure, S, is cover with equal size (ϵ) boxes and an index is seek which can be thought of as the average probability density in each box, defined as $\mu(box(\epsilon))/\epsilon^E$ in a Euclidean space of dimension E. For fractals, instead of density, one speaks in terms of the coarse *Hölder* exponent [12]:

$$\alpha = \frac{\log \mu(box(\epsilon))}{\log \varepsilon}. \tag{1}$$

For a multifractal, α will be restricted to an interval $\alpha_{\min} < \alpha < \alpha_{\max}$, $[\alpha_{\min}, \alpha_{\max}] = I_\alpha$, while for a strictly self-similar fractal there will be a unique α. To obtain the frequency distribution $f(\alpha)$, one must count the number $N(\alpha)$ boxes of size ϵ that have a coarse *Hölder* exponent α (i.e. α in one of the intervals of equal length $\Delta\alpha$ which cover I_α) and take the quotient

$$f(\alpha) = \frac{\log(N(\alpha))}{\log(1/\varepsilon)}. \tag{2}$$

The frequency $f(\alpha)$ obtained by this method will be named $f^*(\alpha)$ from now on.
Another way to compute the $(\alpha, f(\alpha))$ spectrum is using a Legendre coordinate set: a partition function, or a structure function

$$Z(q, \varepsilon) = \sum_i \mu_i^q(\varepsilon) \sim \varepsilon^{\tau(q)} \tag{3}$$

ε as small as possible, is computed. Both functions, $f(\alpha)$ and $\tau(q)$ describe the same aspects of a multifractal, and they are related to each other. In fact, the relationships are

$$\tau(q) = f(\alpha) - q\alpha, \tag{4}$$

where α is given as a function of q by the solution of the equation

$$\frac{d}{d\alpha}(q\alpha - f(\alpha)) = 0. \tag{5}$$

where q is an internal parameter. These last two equations represent the Legendre transform from the variables q and τ to the variables α and $f(\alpha)$.
These two different ways of computing the multifractal spectrum $(\alpha, f^*(\alpha))$ and $(\alpha, f(\alpha))$ will be used in this Section in order to get a sort of taxonomy of the *prior-to-crash* time in the stock market. The spectrum $(\alpha, f^*(\alpha))$ will be referred as *spectrum-by-definition* and $(\alpha, f(\alpha))$ as *Legendre-coordinate-spectrum* throughout this paper.
The value points of the graph of the daily fluctuation of the price at the close fluctuations for the NASDAQ100 composite index as a function of time from 1000 days before the April crash (see Fig. 1) (unit time=1 day), are projected on the vertical axis. These projected points form a Cantor-dust (endowed with the natural measure μ above defined) which is covered with boxes of equal size in order to obtain $f(\alpha)$ and $f^*(\alpha)$. The higher the point in this vertical axis, the higher

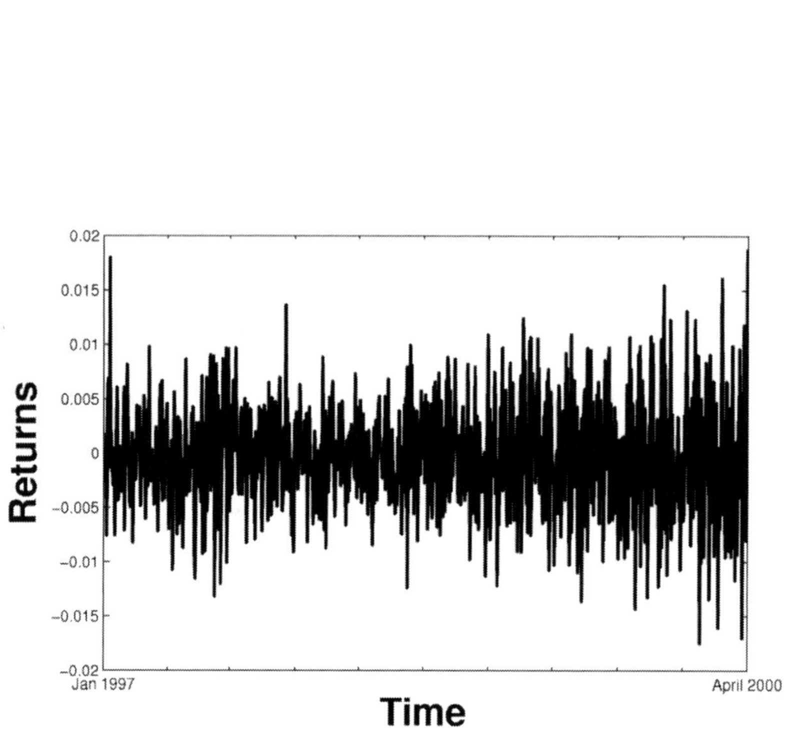

Figure 1. The return time series from 1st Jan 1997 to 10th April 2000 for NASDAQ100 composite index [4].

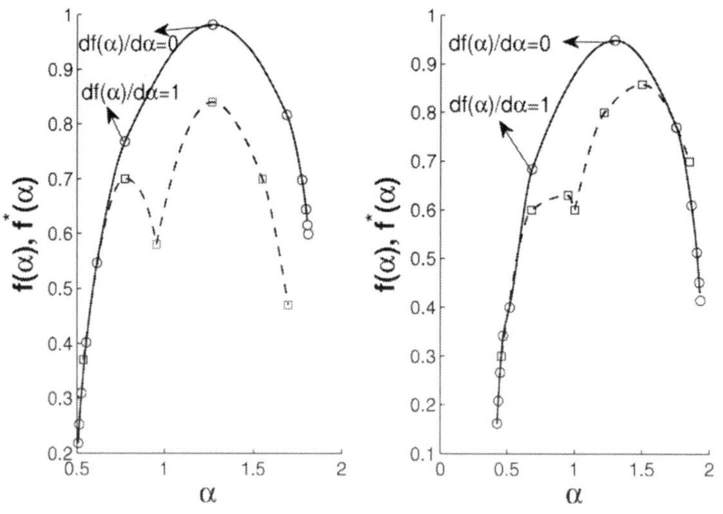

Figure 2. The normalized *spectra-by-definition* $f^*(\alpha)$ in dotted line, superimposed on the *Legendre-coordinate-spectra* $f(\alpha)$, are calculated with a careful analysis of the histogram of I_α. They have slightly shorter domains and lower values than the *Legendre-coordinate-spectra* , due to technical reasons beyond the scope of this paper. Fig 1(left): 1000 days prior to April 2000 NASDAQ100 crash. Fig 1(right): 1600 days prior to April 2000 NASDAQ100 crash, using as many boxes as the squared root of the days analyzed.

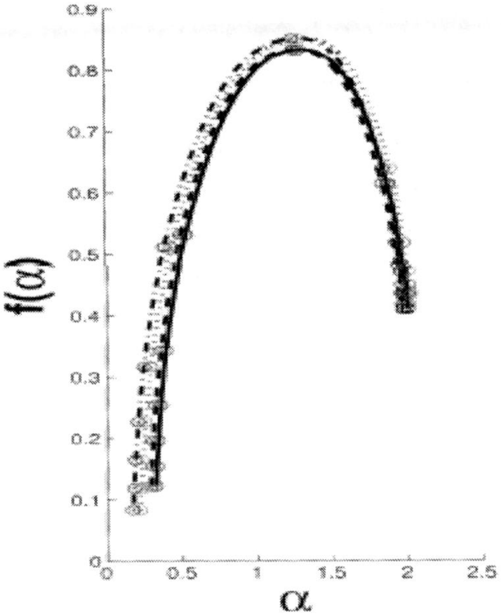

Figure 3. The $f(\alpha)$ for 900 (straight line), 800 (dashed line), 700 (dotted line) and 600 (dash-dot line) days prior to April 2000 NASDAQ100 composite index crash, using as many boxes as the squared root of the days analyzed; the similarity of the shape reflects the fractal structure of the crash point.

the fluctuation of the daily fluctuation of the price at the close is. The *prior-to-crash* situation can be described by observing both *Legendre-corrdinate-spectra* in Fig. 2, which correspond to 1000 and 1600 days prior to crash and relating them with the characteristics mentioned in [3]. These characteristics are about the symmetry of the *Legendre-coordinate-spectrum*, the first one, named *dispersion*, is a very high set of values near α_{max}, that is because there are several points *dispersed* from the fractal process, the second one is the *wide gap* between $df(\alpha)/d\alpha = 1$ and $df(\alpha)/d\alpha = 0$, this characteristic presented in the *Legendre-coordinate-spectrum* is related with the *spectrum-by-definition*, as each case (see Fig 2) the $f(\alpha)$ is the upper one with negative $f''(\alpha)$. The *spectrum-by-definition* shows some qualitative differences: in the *wide gap* between $df(\alpha)/d\alpha = 1$ and $df(\alpha)/d\alpha = 0$ it is observed that $f^*(\alpha)$ decreases, clearly indicating two such $f^*(\alpha)$ colliding in that *wide gap*, which happens when two different measures are processed with the same multifractal algorithm [9] and it was shown in some experiments [11]. In such cases, since $d^2f(\alpha)/d\alpha^2 < 0$, $f(\alpha)$ acts as a smooth upper "roof" for both colliding $f^*(\alpha)$. Poor samples such as our 1000 and 1600 values always yield $f^*(\alpha)$ values smaller than $f(\alpha)$, which can be inferred from the definitions of $f(\alpha)$ and $f^*(\alpha)$ above. The same set of features is shown in Fig. 3.b for smaller samples (900, 800, 700, and 600 days prior to crash). Notice that in this figure the *dispersion* near $\alpha_{\max}$ is larger for 600 days and the similarity of the shape of these spectra suggests the fractal structure of the crash point.

3. The instability quotient

This coefficient is proposed in order to interpret the *wide gap* presented in the multifractal *Legendre-coordinate-spectrum* and the nonconcavity of the *Spectrum-by-definition* in simple time series meaning. The Instability Quotient Q_i measure the number of consecutive fluctuation increment Δp departing from a certain level fluctuation p and not exiting an upper pre-established barrier fluctuation p', normalized to the average of days during which such increase Δp takes place,

$$Q_i = \frac{\text{Number of boxes in a } \Delta\alpha_r \text{ interval}}{\text{Average of days in the } \Delta\alpha_r} \tag{6}$$

Let be p_i the probability measure over the boxes of length l which the *Cantor-like dust*, defined for the NASDAQ 100 composite index in the Introduction, was paved, $\delta_{i,l}$ is the number of points enclosed within the $i-th$ box of length l and T the total number of point belonging the Cantor-like dust

$$p_i = \frac{\delta_{i,l}}{T} \tag{7}$$

taking logarithm in both member of equation 7 and doing some simple algebra

$$\log(p_i) = \frac{\log(\delta_{i,l})}{\log(T)} \Rightarrow \log(p_i) = \log(\delta_{i,l}) - \log(T) \tag{8}$$

it can be noted from equation 8 that $-\log(T)$ is merely a translation and p_i goes as $\delta_{i,l}$, so it is sufficient compute the average of the later in the $r-th$ $\Delta\alpha$ in order to obtain the numerator of Q_i (Eq. 6). The number of boxes in each $\Delta\alpha$ must be counted in order to obtain the denominator of 6 For instance, boxes with days between 78 and 154 belong to the same $\Delta\alpha$ of length $\log(2)$. Their average is slightly over three and a half months (114 days) and we have 4 of these boxes. They are the first $\Delta\alpha$, corresponding to $\alpha_{\min}$. So the corresponding $Q_i = 4/114 = 0.035$. The second $\Delta\alpha$ corresponds to boxes between 31 and 64 days. They also belong to an interval $\Delta\alpha$ of length approx $\log(2)$ and there are 7 such boxes, with an average of 1.5 months (47 days. So $\mathrm{Q_i} = 7/47 = 0.148 \ldots$ and so on (see Table 1). When Q_i is small, e.g. 0.035, it is because fluctuations are unchanged during a long period (3.5 months) and only then go up: this situation is rather stable. When Q_i increases, e.g. 0.148, it is because fluctuations go up every month and a half and they do so no less than seven times: this situation is much more unstable. In order to analyze the behavior of the Q_i, a comparison with the classical binomial, or Moran-like fractal, (see Apendix A) model is done Fig. 4(right), in that figure it is shown Q_i plotted against α for the classical prototype self-similar measure [12] showing always a curvature of the same sign, $d^2\mathrm{Q_i}/d\alpha^2 > 0$, whereas the *prior-to-crash* Q_i corresponding to Fig 4(left) has an inflexion point: a change of sign in $d^2\mathrm{Q_i}/d\alpha^2$. This abnormality is due to an anomaly in the evolution of the market: 5 consecutive boxes with an average of 3 weeks are immediately followed by 7 higher boxes (higher fluctuations) of about 10 days each: an abnormal jump in fluctuations suffered each 10 days (instead of three weeks) for 7 consecutive fluctuation increases. This inflection point takes place in the *wide gap* between $f'(\alpha) = 1$ and $f'(\alpha) = 0$.

4. Conclusions

In this contribution a first step to the developing taxonomy for prior to crash market situation based in the fluctuations of the market value tim series is presented. The huge fluctuations mentioned in [3] are characterized in terms of the so called *dispersion* in Section 3 and the *wide gap* (Section 3) in combination with the *instability index* 4. Regarding the *dispersion*, it means that there are some isolated points in boxes at the very extremes of the vertical *Cantor dust*, defined en Section 2, corresponding to excessive upper jumps in the fluctuation of the returns in Fig. 1 and represented in the *Legendre-coordinate-spectrum* as α_{max} (NB that in Fig 3.left α_{max} has a much higher value than in Fig. 3.right). The other studied characteristic within the *Legendre-coordinate-spectrum* is the *wide gap* presented between ${}^{df(\alpha)}/_{d\alpha} = 1$ and ${}^{df(\alpha)}/_{d\alpha} = 0$ in the *Legendre-coordinate-spectrum*, or the two colliding spectra in the *spectrum-by-definition*,

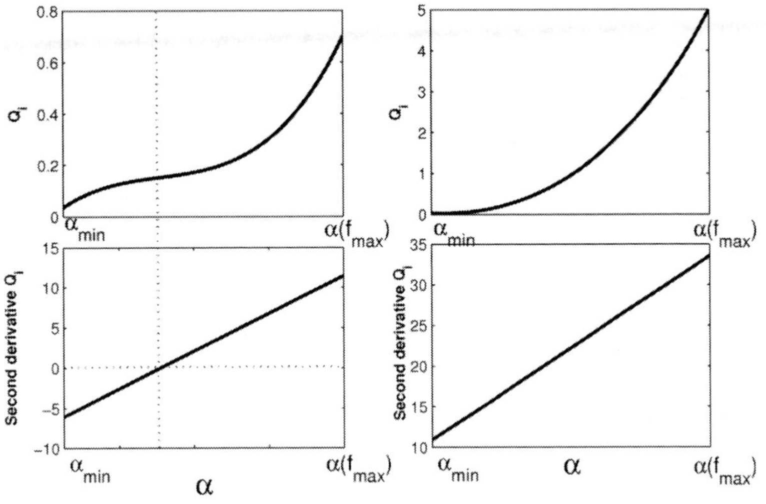

Figure 4. Q_i and its second derivatives computed using a standard finite difference scheme for the NASDAQ100 set (Left column) and for the binomial model [12] (Right column). The inflection point is only presented in the NASDAQ100 case. $I = [\alpha_{min} = 0.6, \alpha_{f_{\max}} = 1.9]$

between the same values. This *wide gap* could be interpreted as a rupture of the multifractal process in the presence of the crash date, this phenomenon is related with some other problems in physics where a phase transition is present or the process itself change its nature due an external force [11]. Q_i function has an inflection point (change of curvature sign) (see Fig. 4.left) and this inflection point is presented in the same α where there the *spectrum-by-definition* lost its concavity, for that reason this problem is related with the non-concave entropy problem studied by [10]. Non-crash situations do not show a change of curvature sign between $\alpha_{\min}$ and $\alpha_{f_{\max}}$, moreover, Fig. 3.a is very symmetric. Moran-like fractals do not shown an inflection point if Q_i function. The inflection point in the Q_i function reflects an instability in the fractal process. This instability is produced in this way: there is a fluctuation increase Δp each 3 weeks, 5 consecutive times, followed by a group of the same increase each 10 days only, and 7 consecutive times: a strong anomaly in the evolution of the market; in terms of Q_i that is the denominators shrinks with a constant near 2 (see Table 1) but the numerator is not growing in the whole interval, instead of that it decreases in the third $\Delta\alpha$, that is exactly where the wide gap is presented in the *Legendre-coordinate-spectrum*. A very large number have been expected in the third and fourth $\Delta\alpha$ in order to get an harmonious Q_i evolution like in a theoretical fractal, see Fig 4. Hence, the inflection point of Q_i function, the *widegap* in the *Legendre-coordinate-spectrum* and the nonconcavity in the *spectrum-by-definition* are in the same $\Delta\alpha$. This gap presented in the *Legendre-coordinate-spetrum* and the two bumps in the spectrum-by-definition could be understand as a fingerprint of the two overlapped carton set proposed by Chakrabarti-Stinchcombre model and this overlap can be measure througth the Instability Quotient, since it doe not present any inflexion point in the presence of only one fractal even if it has two measures (Fig.4).

In summary, a geometrical description of the huge fluctuations mentioned in [3] using fractal geometry and a new index is presented relating both multifractal spectra and the Chakrabarti-Stinchcombre model.

Table 1. This table shows the calculation of each Q_i. The denominators shrink with a constant near 2 but the numerator is not growing in the whole interval, instead of that it decreases in the third $\Delta\alpha$, that is exactly where the wide gap is presented in the *Legendre-coordinate-spectrum.* A very large number have been expected in the third and fourth $\Delta\alpha$ in order to get an harmonious Q_i evolution like in a theoretical fractal, see Fig 4.

Box number, i	Number of boxes in a $\Delta\alpha_i$ interval	Average of days in the $\Delta\alpha_i$
1	4	3.2 *months*
2	7	1.5 *months*
3	5	0.7 *months*
4	7	0.3 *months*

Appendix A

In this appendix a description of the moran-like fractal used in section 3 is done, a more detalied explanation of this kind of fractals can be found in [12]. The unit segment as partitioned in two segments of equal length is considered when $N = 1, \ldots$ and in 2^N equal parts in step N. The first two segments have probability measures p_1 and p_2, $p_1+p_2 = 1$, so p_j, $j = 1, 2$ are contractors .In step N, there are 2^N segments in the unit interval, and their generic probability is ${p_1}^r {p_2}^{N-r}$, r an integer, $0 \leq r \leq N$. The number of segments with this probability is $\binom{N}{r}$. A calculation of its multifractal spectrum $(\alpha, f(\alpha))$ can be done as in [11] in terms of a key frequency $\lambda = \frac{r}{N}$ as internal coordinate, the multifractal spetrum is $\alpha(\lambda) = \frac{-1}{\log 2}(\lambda \log p_1 + (1 - \lambda) \log p_2)$ and $f(\alpha = \frac{-1}{\log 2}\{\lambda \log \lambda + (1 - \lambda) \log(1 - \lambda)\}$

Acknowledgments

F.O.R. very thanks THIC-APFA7 Organizing Comitee for financial support to assist the meeting and Marcel Ausloos for helpfully discussion about this paper and for support before, after and during THIC-APFA7 Conference. F.O.R. thanks Universidad Católica Argentina for its support.

References

[1] Mandelbrot, B.B., Int. Econ. Rev., **1** (1960) 79-106.
[2] Mandelbrot, B.B., The Journal of Business of the Univ. of Chicago, **36** (1963) 394-419.
[3] Vandewalle, N., Ausloos, M., Boveroux Ph. and Minguet A. Eur. Phys. J. B. **9**, (1999) 355-359.
[4] Rotundo, G, in The logistic map and the Route to Chaos, (Springer, The Netherlands, 2006)
[5] Mantegna, R and Stanley E, *An Introduction to Econophysics*, (Cambridge University Press, UK, 2004)
[6] Chakrabarti, B. K. and Stinchcombre R. B., Physica A **207** (1999) 27-34
[7] Chakrabarti B. K., Chatterjee A., Bhattacharyya P (2006) in Takayasu H (Ed) Practical Fruits of Econophysisc, Springer, Tokyo. pp. 107-110.
[8] Y. Huang, A. Johansen, M.W. Lee, H Saluer, D. Sornette: J. Geophys. Res. **105**, 25451 (2000)
[9] Radon, G. and Stoop, R., J. Stat. Phys, **82** 163-180.
[10] Cawley R. and Mauldin, R. D., Adv. Math. **2** (1992) 196-236.
[11] M. Rosen and M. Piacquadio, arXiv:0804.3426.
[12] Falconer, K., *Fractal Geometry* (J. Wiley and Sons, 2003).

Random matrix theory filters and currency portfolio optimisation

J Daly, M Crane and H J Ruskin

Centre for Scientific Computing and Complex Systems Modelling, Dublin City University, Dublin 9, Ireland

E-mail: jdaly@computing.dcu.ie

Abstract. Random matrix theory (RMT) filters have recently been shown to improve the optimisation of financial portfolios. This paper studies the effect of three RMT filters on realised portfolio risk, using *bootstrap* analysis and *out-of-sample* testing. We considered the case of a foreign exchange and commodity portfolio, weighted towards foreign exchange, and consisting of 39 assets. This was intended to test the limits of RMT filtering, which is more obviously applicable to portfolios with larger numbers of assets. We considered both equally and exponentially weighted covariance matrices, and observed that, despite the small number of assets involved, RMT filters reduced risk in a way that was consistent with a much larger S&P 500 portfolio. The exponential weightings indicated showed good consistency with the value suggested by Riskmetrics, in contrast to previous results involving stocks. This decay factor, along with the low number of past moves preferred in the filtered, equally weighted case, displayed a trend towards models which were reactive to recent market changes. On testing portfolios with *fewer* assets, RMT filtering provided *less or no overall risk reduction*. In particular, no long term out-of-sample risk reduction was observed for a portfolio consisting of 15 major currencies and commodities.

1. Introduction

Markowitz portfolio theory [1], an intrinsic part of modern financial analysis, relies on the covariance matrix of returns and this can be difficult to estimate. For example, for a time series of length T, a portfolio of N assets requires $(N^2 + N)/2$ covariances to be estimated from NT returns. This results in estimation noise, since the availability of historical information is limited. Moreover, it is commonly accepted that financial covariances are not fixed over time (e.g. [2–4]) and thus older historical data, even if available, can lead to cumulative noise effects.

Random matrix theory (RMT), first developed by authors such as Dyson and Mehta [5–8], to explain the energy levels of complex nuclei [9], has recently been applied (by several authors including Plerou et al. [9–13] and Laloux et al. [14, 15]) to noise filtering in financial time series, particularly in large dimensional systems such as stock markets. Both groups have analysed US stock markets and have found that the eigenvalues, of the correlation matrix of returns, were consistent with those calculated using random returns, with the exception of a few large eigenvalues. Moreover, their findings indicated that these large eigenvalues, which did not conform to random returns, had eigenvectors that were *more stable* over time [1]. Of

[1] and thus forecasts made here with matrices which have been filtered (i.e. by keeping the stable eigenvalues and smoothing the unstable ones) are expected to show greater reliability over time.

particular interest was the demonstration [9,15], that filtering techniques, based on RMT, could be beneficial in portfolio optimisation, both reducing the realised risk of optimised portfolios, and improving the forecast of this realised risk.

More recently, Pafka et al. [16] extended RMT to provide Riskmetrics type [4] covariance forecasts. Riskmetrics, dating from the 1990's and considered a benchmark in Risk management [16], uses an exponential weighting scheme to model the heteroskedasticity of financial returns. Pafka et al. [16] showed that RMT-based eigenvalue filters can improve the optimisation of minimum risk portfolios, generated using exponentially weighted forecasts. However, these authors found that the decay factors, which produced the least risky portfolios, were higher than the range suggested by Riskmetrics and further concluded that unfiltered Riskmetrics-recommended forecasts were unsuitable for their portfolio optimisation problem.

A recent paper by Sharifi et al. [17], using equally weighted, high frequency returns for estimating covariances, proposed an alternative eigenvalue-filtering method, based on a principal components technique developed by Krzanowski [18] for measuring the stability of eigenvectors, in relation to small perturbations in the corresponding eigenvalues. Sharifi et al. [17] concluded that filtering correlation matrices according to the method outlined in Laloux et al. [15] had a negative effect on this stability. Following this, Daly et al. [19], extended the filtering method of Sharifi et al. [17], and showed that the extended filter offered improvements in terms of risk and stability compared to other RMT filters tested. Indeed, all three RMT filters examined were found, overall, to reduce mean realised risk in all out-of-sample tests. However, in some individual years this was not the case, while on specific days, and for all types of filters, RMT filtering was found to be capable of increasing realised risk, substantially in some cases. In agreement with Pafka et al. [16], the optimal out-of-sample decay factors, for both filtered and unfiltered forecasts, were found to be higher in all cases than those suggested by Riskmetrics [4].

While RMT has been extensively applied to large portfolios, such as the S&P 500, some authors such as Conlon et al. [20], have examined its application to smaller portfolios. Conlon et al. [20] studied a portfolio consisting of 49 hedge funds, with limited historical data, and found that the eigenvalues deviating from RMT contained distinct strategy clustering, while RMT filtering of the correlation matrix resulted in a 35% improvement between the risk of the predicted and realised portfolios.

Our objective in this paper is to evaluate the three RMT filters of Laloux et al. [15], Plerou et al. [9], and Daly et al. [19], applied to both equally and exponentially weighted covariance forecasts, of a trading portfolio consisting of 39 currency pairs and commodities. The effect of filtering on the realised risk of minimum risk portfolios, constructed from these assets, is examined. The filters are studied both in-sample using bootstrapping, and out-of-sample using forward validation, following Daly et al. [19], as described in Sections 3 and 4. Finally, the impact of a further reduction in asset numbers is also assessed.

The paper is organised as follows. Section 2 reviews the theoretical background for the three RMT filters. Section 3 contains an in-sample analysis of the filters, including an analysis of reducing the number of assets from 432, as studied in Daly et al. [19], to 39, focusing on S&P data. Section 4 contains the results, of the out-of-sample tests, on effectiveness of the filters in reducing realised risk. In Section 5, the impact of further asset number reduction is studied, and Section 6 contains the concluding remarks.

2. Background

2.1. Random Matrix Theory and Historical Covariance

As described by Laloux et al. [14], Plerou et al. [9], Sharifi et al. [17] and others, in the context of correlation matrices of financial returns, if $\mathbf{R}$ is any matrix defined by

$$\mathbf{R} = \frac{1}{T}\mathbf{A}\mathbf{A}' \tag{1}$$

where $\mathbf{A}$ is an $N \times T$ matrix whose elements are i.i.d. [2] random variables with zero mean, then it has been shown [21] that, in the limit $N \to \infty, T \to \infty$ such that $Q = T/N \geq 1$ is fixed, the distribution $P(\lambda)$ of the eigenvalues of $\mathbf{R}$ is self-averaging, and is given by

$$P(\lambda) = \begin{cases} \frac{Q}{2\pi\sigma^2} \frac{\sqrt{(\lambda_+ - \lambda)(\lambda - \lambda_-)}}{\lambda} & \text{if } \lambda_- \leq \lambda \leq \lambda_+ \\ 0 & \text{otherwise} \end{cases} \tag{2}$$

where σ^2 is the variance of the elements of $\mathbf{A}$ and

$$\lambda_\pm = \sigma^2 \left(1 + 1/Q \pm 2\sqrt{1/Q}\right). \tag{3}$$

Financial correlation and covariance matrices can be expressed, in general, in the form given by Equation (1), so matrices for historical data can be compared to those generated from random returns. Here we define the *covariance* matrix $\mathbf{V} = \{\sigma_{ij}\}_{i,j=1}^N$ of returns [3] by

$$\sigma_{ij} = \langle G_i(t) G_j(t) \rangle - \langle G_i(t) \rangle \langle G_j(t) \rangle \tag{4}$$

where $\langle \cdot \rangle$ refers to the mean over time, and the *correlation* matrix $\mathbf{C} = \{\rho_{ij}\}_{i,j=1}^N$ is given by

$$\rho_{ij} = \sigma_{ij} / \sqrt{\sigma_{ii}\sigma_{jj}} \tag{5}$$

where $\{G_i(t)\}_{t=1,\ldots,T}^{i=1,\ldots,N}$ are the returns

$$G_i(t) = \ln(S_i(t)/S_i(t-1)) \tag{6}$$

and where $S_i(t)$ is the spot price of asset i at time t.

2.2. Random Matrix Theory and Exponentially Weighted Covariance

In extending RMT filtering to exponentially weighted matrices, Pafka et al. [16] have analysed matrices of the form $\mathbf{M} = \{m_{ij}\}_{i,j=1}^N$ with

$$m_{ij} = \sum_{k=0}^{\infty} (1-\alpha)\alpha^k x_{ik} x_{jk} \tag{7}$$

and where $\{x_{ik}\}_{k=0,\ldots,\infty}^{i=1,\ldots,N}$ are assumed to be $N.I.D.(0, \sigma^2)$. [4] They have shown that, in the special case $N \to \infty$, $\alpha \to 1$ with $Q = 1/(N(1-\alpha))$ fixed, the density, $\rho(\lambda)$, of the eigenvalues of $\mathbf{M}$ is given by $\rho(\lambda) = Qv/\pi$ where v is the root of

$$F(v) = \frac{\lambda}{\sigma^2} - \frac{v\lambda}{\tan(v\lambda)} + \ln(v\sigma^2) - \ln(\sin(v\lambda)) - \frac{1}{Q} \tag{8}$$

$F(v)$ is well defined on the open interval $(0, \pi/\lambda)$. If a root does not exist on this interval for a given value of λ we define $\rho(\lambda) = 0$ for that λ. The family of matrices, defined by Equation (7), includes the Riskmetrics [4] covariance and correlation matrices. Following this, we define the exponentially weighted covariance matrix $\mathbf{V}^* = \{\sigma_{ij}^*\}_{i,j=1}^N$ by

[2] i.i.d. ≡ independent and identically distributed

[3] throughout this paper the following notation is used: $\{x_i\}_{i=1}^N \equiv \{x_i : i = 1, \ldots, N\}$, $\{x_{ij}\}_{i,j=1}^N \equiv \{x_{ij} : i = 1, \ldots, N; j = 1, \ldots, N\}$, $\{x_{it}\}_{t=1,\ldots,T}^{i=1,\ldots,N} \equiv \{x_{it} : i = 1, \ldots, N; t = 1, \ldots, T\}$ etc.

[4] $N.I.D.(\mu, \sigma^2)$ ≡ Normally and identically distributed (with mean μ and variance σ^2)

$$\sigma_{ij}^* = \frac{1-\alpha}{1-\alpha^T} \sum_{t=0}^{T-1} \alpha^t (G_i(T-t) - \langle G_i \rangle)(G_j(T-t) - \langle G_j \rangle) \tag{9}$$

and define the corresponding, exponentially weighted, correlation matrix $\mathbf{C}^* = \{\rho_{ij}^*\}_{i,j=1}^N$ by

$$\rho_{ij}^* = \sigma_{ij}^* / \sqrt{\sigma_{ii}^* \sigma_{jj}^*} \tag{10}$$

Here, α is commonly called the *decay factor*.

The maximum eigenvalue of an exponentially weighted random matrix can be found using Equation (8). Equivalently [19], the maximum eigenvalue is the solution of

$$\frac{\lambda}{\sigma^2} - \ln\left(\frac{\lambda}{\sigma^2}\right) = 1 + \frac{1}{Q}\ , \quad \lambda > \sigma^2 \tag{11}$$

2.3. Krzanowski Stability

One of the filtering methods discussed, Sharifi et al. [17], and considered also here, is based on the stability, as described by Krzanowski [18], of the filtered matrix. Krzanowski [18] measured eigenvector stability, specifically the effect on each eigenvector of a perturbation in the corresponding eigenvalue. This is in contrast to stability over time, as analysed by many other authors, e.g. [9, 15]. Krzanowski [18] considered the angle, θ_i, between an eigenvector v_i and v_i^p, where v_i^p is the maximum perturbation that can be applied to v_i while ensuring that the eigenvalue, λ_i^p, corresponding to v_i^p is within ϵ of the eigenvalue, λ_i, corresponding to v_i, and showed that θ_i is given by:

$$\cos\theta_i = \begin{cases} \left(1 + \frac{\epsilon}{\lambda_i - \lambda_{i-1}}\right)^{-\frac{1}{2}} & \text{for } \lambda_i^p < \lambda_i \\ \\ \left(1 + \frac{\epsilon}{\lambda_{i+1} - \lambda_i}\right)^{-\frac{1}{2}} & \text{for } \lambda_i < \lambda_i^p \end{cases} \tag{12}$$

where $\lambda_1 \leq \lambda_2 \leq \ldots \leq \lambda_L$ are the eigenvalues.

2.4. Filtering Methods

All three filtering methods compared here are based on replacing the "noisy" eigenvalues of the correlation matrix, while maintaining its trace. The noisy eigenvalues are taken to be those that are less than, or equal to, the maximum possible eigenvalue of the corresponding random matrix (as defined in Sections 2.1 and 2.2). The correlation matrix being filtered is first deconstructed via the eigen decomposition theorem [5]. The noisy eigenvalues are subsequently replaced using one of the three methods outlined here, and the matrix is rebuilt, again referring to the eigen decomposition theorem, resulting in the filtered matrix, $\mathbf{M}_{\text{filtered}}$.

The first filtering method examined is that of Laloux et al. [15] (referred to here as LCPB), which replaces the noisy eigenvalues with their mean, thus maintaining the trace. The second filtering method is that described by Plerou et al. [9] (and referred to here as PG+). This method replaces the noisy eigenvalues by zeroes and, after $\mathbf{M}_{\text{filtered}}$ is built, replaces its main diagonal with that of the original matrix $\mathbf{M}$, again preserving the trace. The third filtering method, Daly et al. [19], is adapted from that of Sharifi et al. [17]. To maximise the Krzanowski stability of the filtered matrix, while also maintaining its trace, the method of Sharifi et al. [17] replaces

[5] Let $\mathbf{M}$ be a square matrix and let $\mathbf{E}$ be the matrix of its eigenvectors. If $\mathbf{E}$ is a square matrix then $\mathbf{M} = \mathbf{EDE}^{-1}$ where $\mathbf{D}$ is a diagonal matrix containing the corresponding eigenvalues on the main diagonal. [22]

the noisy eigenvalues with ones that are equally and maximally spaced, are positive, and have sum equal to the sum of those replaced. To achieve maximal spacing, it is assumed that the smallest replacement eigenvalue should be very close to zero. The method described in Daly et al. [19], referred to here as the KR method, adapts this by making the smallest replacement eigenvalue a parameter of the filter. This parameter can then be adjusted to maximise the forecasting potential of the filter. It was found, in Daly et al. [19], that the optimal parameter value for reducing realised portfolio risk involved some reduction in stability. However, the KR method was found to improve stability, compared to both the LCPB and PG+ methods. The non-limiting case KR methods considered here, and their defining minimum replacement eigenvalues, were: KR2 ($\frac{1}{2}\overline{\Lambda}_{\text{noisy}}$), KR4 ($\frac{1}{4}\overline{\Lambda}_{\text{noisy}}$), KR8 ($\frac{1}{8}\overline{\Lambda}_{\text{noisy}}$), KR16 ($\frac{1}{16}\overline{\Lambda}_{\text{noisy}}$), KR64 ($\frac{1}{64}\overline{\Lambda}_{\text{noisy}}$), KR100 ($\frac{1}{100}\overline{\Lambda}_{\text{noisy}}$) and KR1000 ($\frac{1}{1000}\overline{\Lambda}_{\text{noisy}}$), where $\overline{\Lambda}_{\text{noisy}}$ is the mean of the noisy eigenvalues. The limiting minimum replacement eigenvalues, $\overline{\Lambda}_{\text{noisy}}$ and 10^{-8}, were also considered.

2.5. Data

In order to examine the effects on RMT filtering of reducing the number of assets, from that of the S&P portfolio studied in Daly et al. [19], to that of the currency and commodity portfolio analysed here, daily closing price data for the S&P 500 index stocks were used. The index composition was taken as of 1st February 2006[6]. The dataset runs from 1st June 1995 to 1st February 2006, and any series not covering the entire period were discarded, leaving a total of 432 stocks. To examine the application of RMT filters to the currency and commodity portfolio, and for all the remaining analysis, daily currency spot prices vs. the US Dollar (USD) were used, along with equivalent rates for Silver, Gold, Platinum and Oil. This data, provided by Pacific Exchange Rate Service[7], covered the period from 4th January 1999 to 31st December 2007. The currencies and commodities selected for this analysis are outlined in Table 1. This group of assets is referred to as the "Fx portfolio" throughout this paper.

3. In-Sample Analysis

For the in-sample analysis, and following [16], bootstrapped samples were taken, together with the mean across these samples. For a given value of N (the number of assets), we randomly selected N assets from the data set, and a random test date. Everything up to and including the test date was taken as historical information and everything afterwards as realised, future information. For each N, we repeated this random selection 1000 times, with replacement, and calculated the mean, across all bootstrapped samples, of the realised risk of the forecast minimum risk portfolio [16] (calculated using our forecast covariance).

A covariance forecast in this context consisted of a raw forecast, which was either exponentially or equally weighted, and could be unfiltered, or filtered by one of the LCPB, PG+ or KR methods applied to the correlation matrix. Filtering the covariance matrix directly, along the lines of Daly et al. [19], was found to be unsuitable when applied to the Fx portfolio, increasing out-of-sample mean realised risk, in most cases, compared to the correlation filter.

On each test date, we calculated the forecast minimum risk portfolio, optimised as follows [16]. Choose a portfolio weighting $\{w_i\}_{i=1}^{N}$ that minimises

$$\sum_{i,j=1}^{N} w_i w_j \hat{\sigma}_{ij} \tag{13}$$

[6] from www.standardandpoors.com

[7] http://fx.sauder.ubc.ca/data.html

Table 1. List of currencies and commodities used. The base currency was the U.S. Dollar.

Name	Code	Name	Code
Australian Dollar	AUD	Peruvian Nuevo Sol	PEN
Brazilian Real	BRL	Philippine Peso	PHP
Canadian Dollar	CAD	Platinum Ounce	XPT
Chilean Peso	CLP	Polish Zloty	PLN
Colombian Peso	COP	Romanian New Leu	RON
Czech Koruna	CZK	Russian Rouble	RUB
Euro	EUR	Silver Ounce	XAG
Fijian Dollar	FJD	Singapore Dollar	SGD
Gold Ounce	XAU	Slovak Koruna	SKK
Hungarian Forint	HUF	South African Rand	ZAR
Icelandic Krona	ISK	South Korean Won	KRW
Indian Rupee	INR	Sri Lankan Rupee	LKR
Indonesian Rupiah	IDR	Pound Sterling	GBP
Israeli New Sheqel	ILS	Swedish Krona	SEK
Japanese Yen	JPY	Swiss Franc	CHF
Mexican Peso	MXN	Thai Baht	THB
Moroccan Dirham	MAD	Thaiwanese Dollar	TWD
New Zealand Dollar	NZD	Tunisian Dinar	TND
Norwegian Krone	NOK	Turkish New Lira	TRY
Oil, Brent Crude, Barrel	XCB		

while satisfying the budget constraint

$$\sum_{i=1}^{N} w_i = 1 \tag{14}$$

where $\hat{\mathbf{V}} = \{\hat{\sigma}_{ij}\}_{i,j=1}^{N}$ is the forecast covariance matrix. The solution, $\{\hat{w}_i\}_{i=1}^{N}$, of this problem is:

$$\hat{w}_i = \frac{\sum_{j=1}^{N} \hat{\sigma}_{ij}^{-1}}{\sum_{j,k=1}^{N} \hat{\sigma}_{jk}^{-1}} \quad \forall i \tag{15}$$

where $\hat{\mathbf{V}}^{-1} = \{\hat{\sigma}_{ij}^{-1}\}_{i,j=1}^{N}$ is the matrix inverse of $\hat{\mathbf{V}}$. The *realised risk* of the optimal portfolio is defined by

$$\sqrt{\sum_{i,j=1}^{N} \hat{w}_i \hat{w}_j \tilde{\sigma}_{ij}} \tag{16}$$

Here, $\tilde{\mathbf{V}} = \{\tilde{\sigma}_{ij}\}_{i,j=1}^{N}$ is the realised covariance matrix, and is the (equally weighted) covariance matrix of the realised future returns over the investment period. We used a forecasting period of 20 days throughout. The *forecast risk* is calculated analogously, using the forecast covariance matrix, $\hat{\mathbf{V}}$.

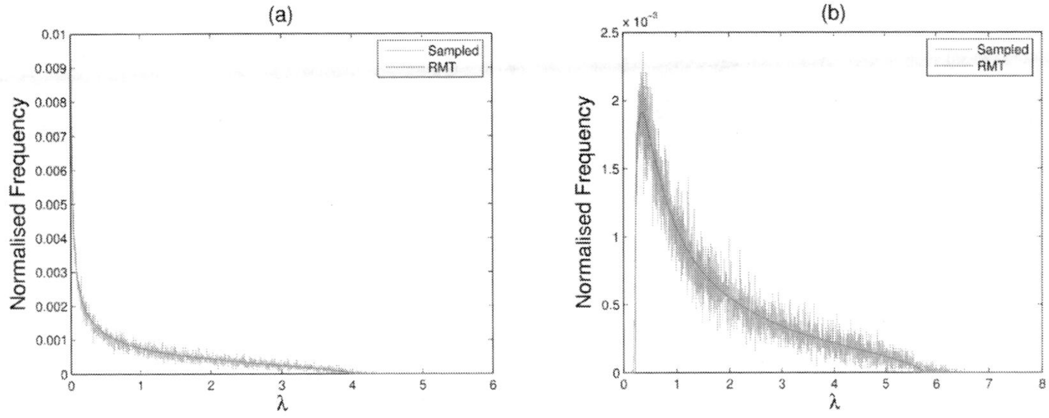

Figure 1. Eigenvalue distribution of 1000 sampled random matrices vs RMT approximations, for 39 assets, with (a) equal weights and $T = 40$ historical data points and (b) exponential weights and a decay factor of $\alpha = 0.97$.

By comparing the covariance forecasts in this way, we measured their effect on realised risk without using forecast returns, which would have introduced additional noise into the results. Further, we have not used any knowledge of future returns in our tests, since we wished to evaluate both forecasting methods (equal vs. exponential weighting) as well as filtering methods. This is in contrast to some previous studies (e.g. [15]), that have isolated the effect of the filtering method on the correlation matrix, by using future knowledge of realised returns to estimate the variance of each individual asset.

3.1. Reduction of S&P Assets

We first analysed the effect on risk of reducing the number of assets in the S&P 500 portfolio [19], from 432 to 39. As a preamble, Figure 1 shows a sample comparison, between the eigenvalue distribution of 1000 sampled random matrices with 39 assets, and the corresponding RMT approximation, for some equally and exponentially weighted random matrices. Here we saw close agreement between the sampled distributions and their RMT approximations. However, the RMT distributions, which are approximations as $N \to \infty$, underestimated the maximum random eigenvalue in both sample cases. Despite this, subsequent results showed that RMT filters could provide risk reduction for portfolios of 39 assets.

Figures 2 and 3 show the effect of RMT filtering on in-sample mean realised risk, for different sizes of S&P portfolio, in the equally and exponentially weighted cases respectively. Here we saw that filtering can be effective for low asset numbers. This was observed in the equally weighted case when the amount of historical data used was low (equivalent to a low value of Q), and in the exponentially weighted case across a wide range of decay factors. The effect of filtering was however seen to be reduced, in general, as the number of assets was decreased. The best unfiltered result also approached the best filtered for lower values of N.

These results also confirmed previous work [23, 24], which showed via simulations that as the value of Q increased, for example through the use of more data, the effect of noise reduced. We also noted another effect here. For the higher values of N, the best overall risk in the equally weighted case was achieved by *reducing* the amount of data *and filtering*, thereby uncovering flavours of models that were previously masked by noise.

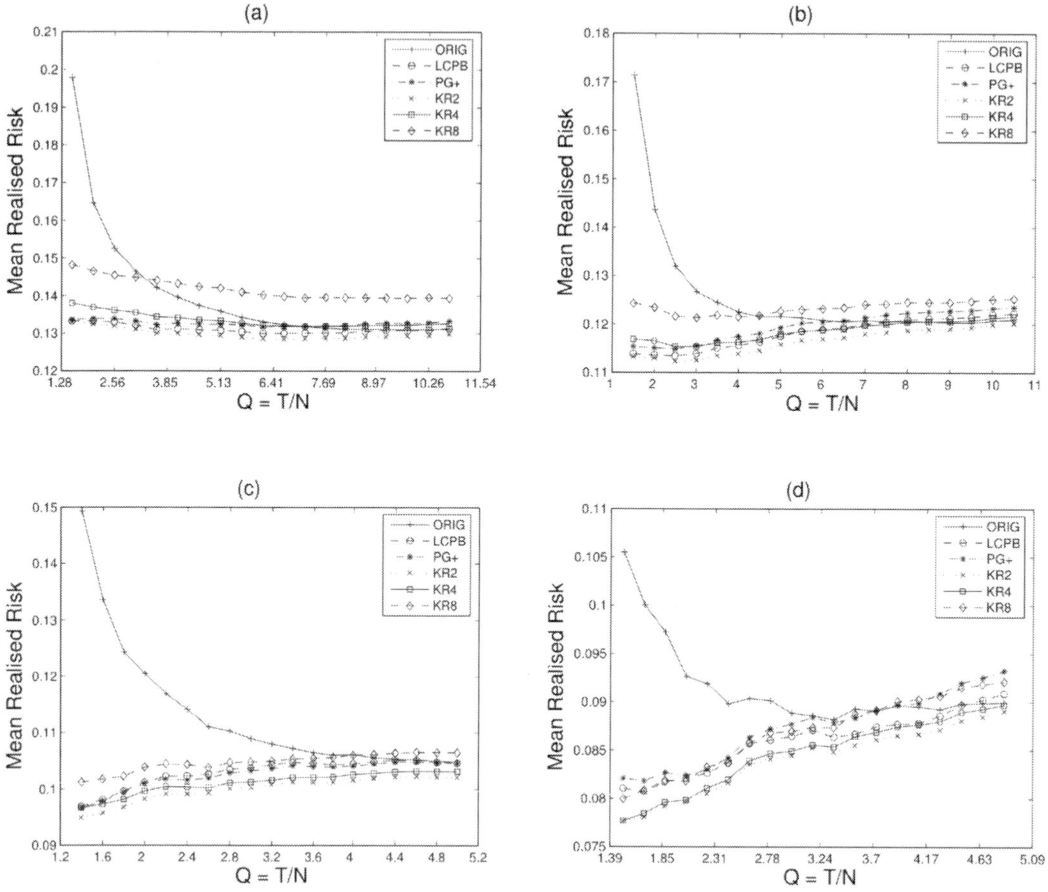

Figure 2. Mean bootstrapped (in-sample) realised risk, for selected filters, applied to equally weighted volatility forecasts of S&P stocks, and for unfiltered volatility ("ORIG"), for N = (a) 39, (b) 100, (c) 250 and (d) 432 assets. The x-axis shows the Q-value, which is the number of past moves used, divided by the number of assets in the portfolio.

3.2. In-Sample Results for the Fx Portfolio

Figure 4 shows the corresponding effect of RMT filtering on the in-sample mean realised risk of the Fx portfolio, with 39 assets. As in the S&P case [19], the in-sample results showed, in general, the potential of RMT filters to reduce realised risk, and the KR2, KR4 and KR8 methods were found to be amongst the best performing of all filters, and were reasonably consistent with each other, while optimisation performance of the KR filters disimproved, in general, as the minimum replacement eigenvalue approached zero. In the equally weighted case, a preference was again shown here for the lowest available number of past moves to be used (equivalent to the lowest Q value), in conjunction with RMT filtering. In the exponentially weighted case, RMT filtering was again preferred overall, and the optimal in-sample decay factor coincided with the Riskmetrics [4] recommendation of 0.97, in contrast to the S&P case [19]. For both weightings, we observed that the use of filtering allowed for a *more reactive model* to be used. In the equally weighted case in particular, good agreement was found with the S&P portfolio. However, the closest agreement was clearly between the Fx portfolio with 39 assets, and the S&P portfolio with 432, and not 39 as might be expected.

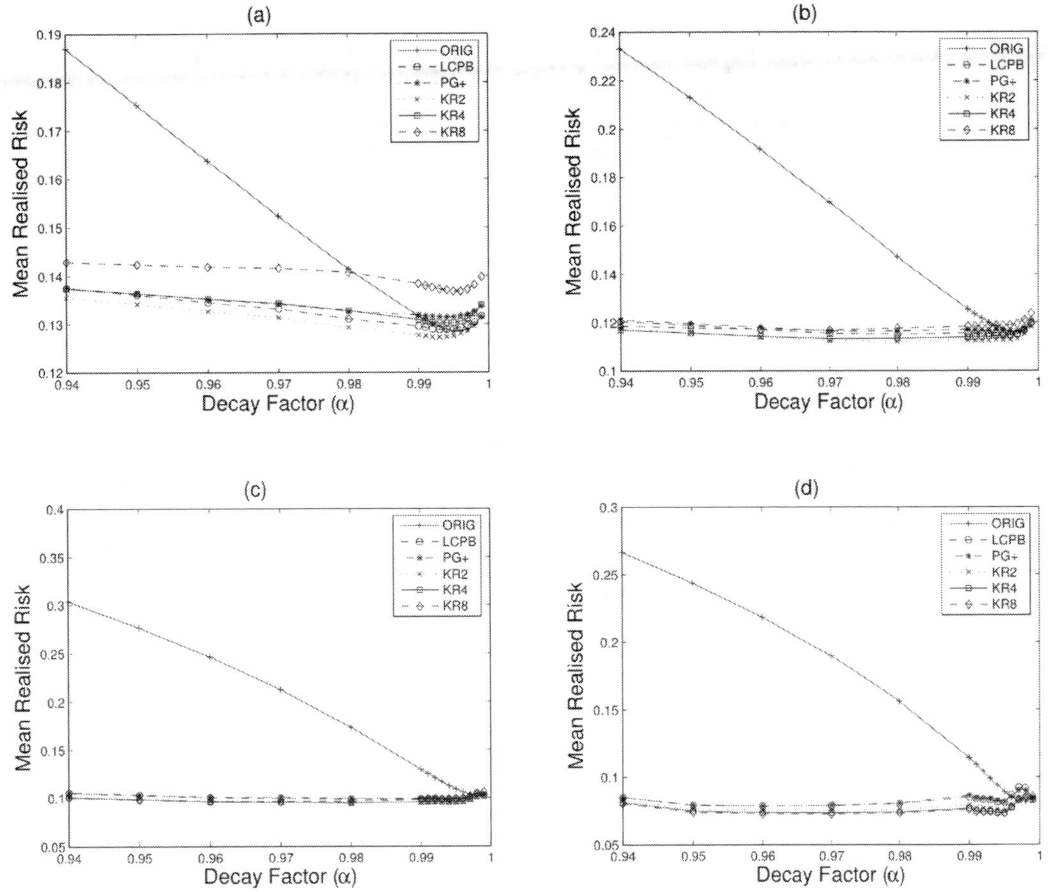

Figure 3. Mean bootstrapped (in-sample) realised risk, for selected filters, applied to exponentially weighted volatility forecasts of S&P stocks, and for unfiltered volatility ("ORIG"), for $N =$ (a) 39, (b) 100, (c) 250 and (d) 432 assets.

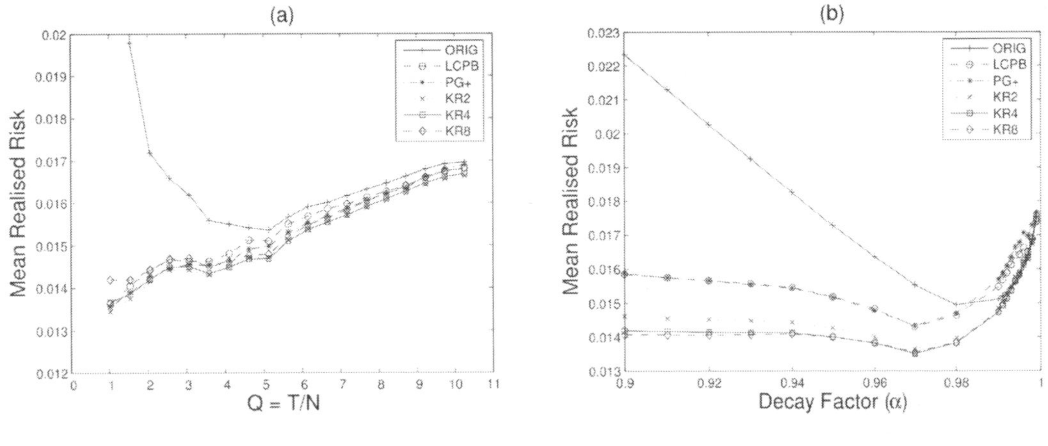

Figure 4. Mean bootstrapped (in-sample) realised risk, for selected filters, applied to (a) equally weighted and (b) exponentially weighted volatility forecasts, and for unfiltered volatility ("ORIG"), for the Fx portfolio, with 39 assets.

4. Out-of-Sample Analysis of the Fx Portfolio

For comparing the models out-of-sample we used forward validation. This method considers every available test date and for each one uses data prior to the test date to optimise any model parameters. This allows the comparison of filtering methods with different numbers of parameters and also gives some insight into the stability of the models over time. The value of the weighting parameter (α or T) and the choice of KR model were determined out-of-sample. Possible KR models were all those defined in Section 2.4, as well as the LCPB model (for completeness). The forward validation was performed over a period of 1837 days, 129 of which were used as the initial training period. Subsequent retraining was done daily. We again used the realised risk of the forecast minimum risk portfolio as our metric and all 39 assets were used to eliminate the need to arbitrarily choose assets each day.

Table 2 shows the performance of the covariance forecasting and filtering combinations. The figures shown are mean realised risk as a percentage of the result for unfiltered equally weighted covariance. RMT filtering was seen *on average* to reduce realised risk in all cases, compared to the unfiltered portfolio. The best overall performance was seen when applying the PG+ filter to the equally weighted case. In the exponentially weighted case, KR filtering was found to be best, which was also the case for the larger S&P 500 portfolio [19], and at similar levels of risk reduction. We further observed that RMT filtering reduced mean realised risk in the majority of individual years. However, while realised risk was reduced on the majority (62.6%) of individual days, the filters were also found capable of *increasing* realised risk substantially on any one day. These results were consistent with recent results for a much larger S&P 500 portfolio [19].

Table 2. Mean out-of-sample realised risk as a percentage of that for unfiltered equally weighted covariance, for the full Fx portfolio, with 39 assets. Filtering was seen to reduce mean realised risk in all cases.

Model	Unfiltered	LCPB	PG+	KR
Equal Weights	100	87	85.8	86.6
Exponential Weights	98.1	91.8	92.7	87.4

The optimal number of past moves (equally weighted), and decay factors, as selected through time by the forward validation, are shown in Figure 5. In this case the decay factors chosen were quite consistent with the value of 0.97 suggested by Riskmetrics [4], unlike in the S&P case [16,19]. In the equally weighted case, low numbers of past moves were ultimately preferred on average, in agreement with the in-sample results. Here, this resulted in equally weighted models outperforming exponential, when filtered. A trend towards models which were reactive to recent market changes can be seen in both cases.

5. Further Asset Reduction

Finally, we examined the effect of filtering an Fx portfolio with a reduced number of assets. Figures 6 and 7 show the effect of RMT filtering on in-sample mean realised risk, for portfolios with $N = 10, 15, 20$ and 30 assets, in the equally and exponentially weighted cases respectively. RMT filtering was seen to be effective for a small Q value, at all four values of N, and so, in an environment where the number of past data points, T, is limited, either by choice or necessity, RMT filtering may be of some benefit. However, without any restrictions on T, the effectiveness of RMT filtering was seen to be reduced as N is reduced. In the case of exponential weights, RMT filtering was seen to provide most benefit for lower decay factors. However, in the presence of a free choice of decay factor, filtering provided little or no overall risk reduction for $N \leq 20$.

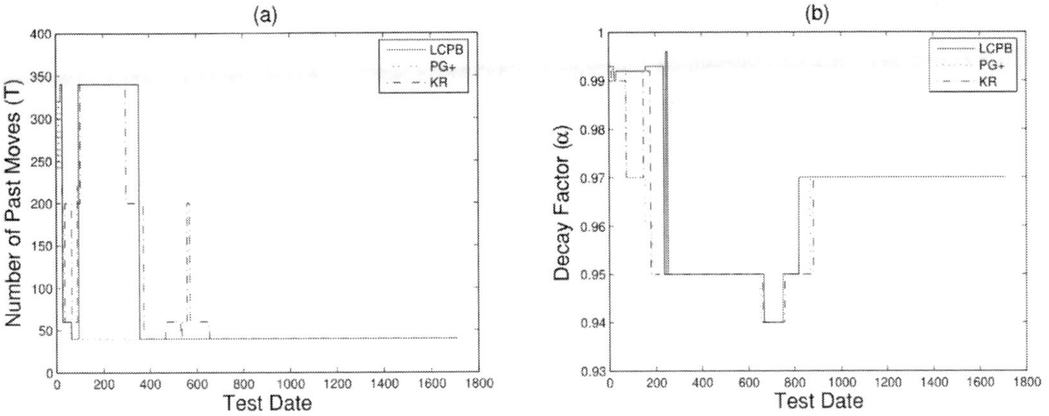

Figure 5. Optimal number of (a) past moves and (b) decay factor values, as selected by forward validation, for the Fx portfolio. In the unfiltered cases (not shown), the number of past moves ranged between 200 and 240, and the decay factors were usually equal to 0.98, and sometimes in the range 0.99 - 0.995.

These results for the Fx portfolio showed agreement with those for the S&P 500, although in the two markets, similar effects of asset reduction occurred at *different values of* N. In both cases, filtering was most effective for larger portfolios.

These conclusions were also reflected in the out-of-sample results for a portfolio consisting of 15 major currencies and commodities, selected from the main portfolio. In this case, RMT filtering was found to provide no benefit in the long run, as shown in Table 3, where filtering was seen to increase risk in most cases.

Table 3. Mean out-of-sample realised risk as a percentage of that for unfiltered equally weighted covariance, for the Fx portfolio with 15 major assets. Filtering was seen to increase mean realised risk in most cases.

Model	Unfiltered	LCPB	PG+	KR
Equal Weights	100	103.8	103.1	99.1
Exponential Weights	97.1	103.1	102.6	98.2

6. Conclusions

We have studied the application of RMT filters to a currency and commodity portfolio consisting of just 39 assets and found that, broadly, our results were in agreement with those reported [16, 19], that RMT-based filtering can improve the realised risk of minimum risk portfolios, despite the *low number of assets* involved in this case.

Using forward validation, RMT filters were found, overall, to reduce mean realised risk in all cases tested, and the majority of years and days. However, they were also found capable of increasing realised risk substantially on some individual days. These results were consistent with the S&P 500 case previously examined [19]. *In contrast to* these previous results, the decay factors chosen here showed good consistency with the value of 0.97 suggested by Riskmetrics [4]. In the equally weighted case, the value of Q ultimately preferred, when filtering, involved the

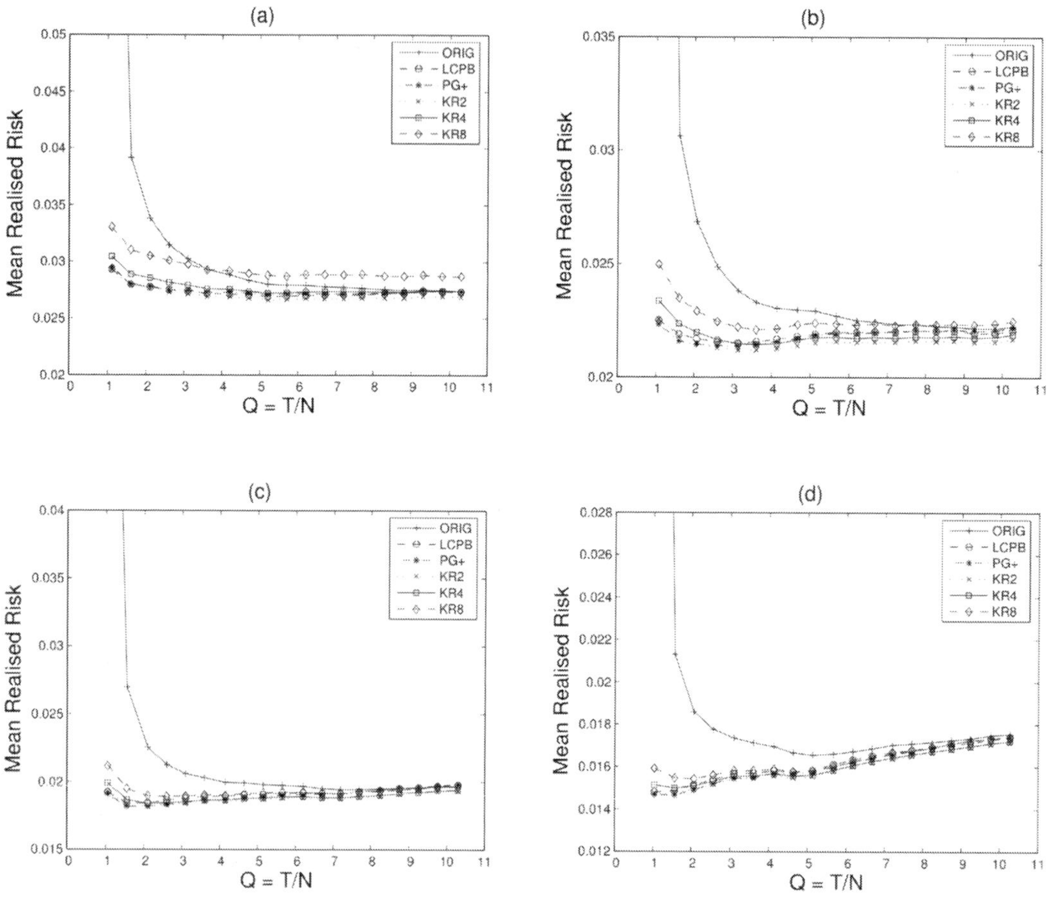

Figure 6. Mean bootstrapped (in-sample) realised risk, for selected filters, applied to equally weighted volatility forecasts of the Fx portfolio, and for unfiltered volatility ("ORIG"), for (a) 10, (b) 15, (c) 20, and (d) 30 assets.

least amount of historical data possible, indicating *very reactive models*. These low values of Q would have been unsuitable in the absence of filtering, and were also seen in the S&P case when the maximum amount of stocks was used.

The observed behaviour of the stability-based filter was in agreement with that of the S&P case [19], namely that the KR2, KR4 and KR8 methods were amongst the best performing of all filters, and reasonably consistent with each other, while optimisation performance disimproved, in general, as the minimum replacement eigenvalue approached zero. In the exponentially weighted case, the KR filter outperformed the others by a factor of 70%.

When RMT filtering was applied to Fx portfolios with fewer asset numbers it was observed, in general, that the benefit of filtering was reduced as asset numbers (N) decreased. Similar reduction was noted for the S&P portfolio, but at different values of N. This effect was also reflected in the out-of-sample filter performance, for a portfolio consisting of 15 major currencies and commodities, where RMT filtering provided no long term risk reduction, and was more likely to *increase* realised risk.

Taken as a whole, these results suggested that RMT filtering can provide strong risk reduction for foreign exchange and commodity portfolios involving sufficient numbers of assets. RMT filters also uncovered different styles of models than were possible with unfiltered analysis, namely ones

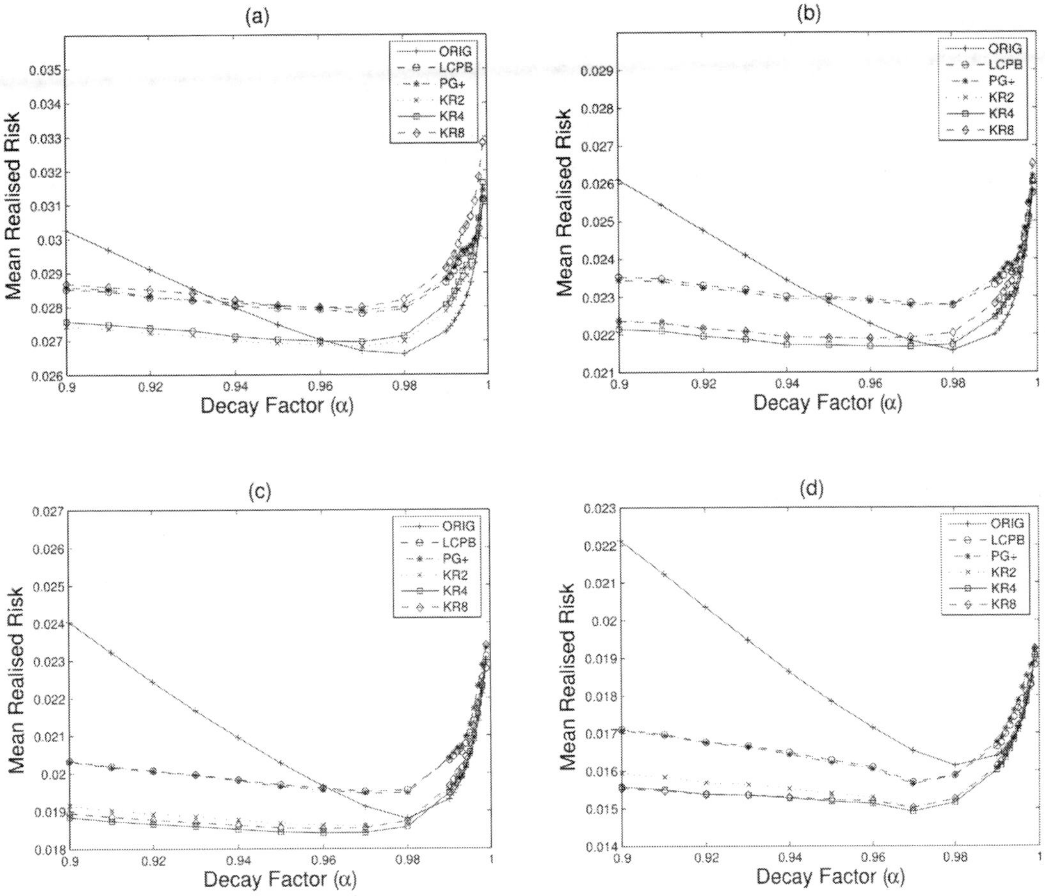

Figure 7. Mean bootstrapped (in-sample) realised risk, for selected filters, applied to exponentially weighted volatility forecasts of the Fx portfolio, and for unfiltered volatility ("ORIG"), for (a) 10, (b) 15, (c) 20, and (d) 30 assets.

that reacted quickly to market events. Without filtering these models utilising very recent data were found to be hidden by noise.

Acknowledgements

We would like to thank the Irish Research Council for Science, Engineering and Technology for supporting this research under the Embark Scholarship Scheme.

References

[1] Markowitz H 1959 *Portfolio Selection: Efficient Diversification of Investments* (New York: Wiley)
[2] Bouchaud J and Potters M 2000 *Theory of Financial Risks - From Statistical Physics to Risk Management* (U.K.: Cambridge University Press)
[3] Dacorogna M, Gençay R, Muller U, Olsen R and Pictet O 2001 *An Introduction to High-Frequency Finance* (U.S.: Academic Press)
[4] J P Morgan and Reuters 1996 Riskmetrics Tech. rep. http://www.riskmetrics.com/techdoc.html
[5] Dyson F 1962 *J. Math. Phys.* **3** 140–75
[6] Dyson F and Mehta M 1963 *J. Math. Phys.* **4** 701–12
[7] Mehta M and Dyson F 1963 *J. Math. Phys.* **4** 713–19
[8] Mehta M 1991 *Random Matrices* (New York: Academic Press)

[9] Plerou V, Gopikrishnan P, Rosenow B, Amaral L, Guhr T and Stanley H 2002 *Phys. Rev.* E **65** 066126
[10] Plerou V, Gopikrishnan P, Rosenow B, Amaral L and Stanley H 1999 *Phys. Rev. Lett.* **83** 1471–74
[11] Plerou V, Gopikrishnan P, Rosenow B, Amaral L and Stanley H 2000 *Physica* A **279** 443–56
[12] Plerou V, Gopikrishnan P, Rosenow B, Amaral L and Stanley H 2000 *Physica* A **287** 374–82
[13] Plerou V, Gopikrishnan P, Rosenow B, Amaral L and Stanley H 2001 *Physica* A **299** 175–80
[14] Laloux L, Cizeau P, Bouchaud J and Potters M 1999 *Phys. Rev. Lett.* **83** 1467–70
[15] Laloux L, Cizeau P, Potters M and Bouchaud J 2000 *Int. J. Theor. Appl. Finance* **3** 391–97
[16] Pafka S, Potters M and Kondor I 2004 Exponential weighting and random-matrix-theory-based filtering of financial covariance matrices for portfolio optimization; To appear in *Quant. Fin.* (*Preprint* `cond-mat/0402573`)
[17] Sharifi S, Crane M, Shamaie A and Ruskin H J 2004 *Physica* A **335** 629–43
[18] Krzanowski W 1984 *J. Royal Stat. Soc.* B **46** 558–63
[19] Daly J, Crane M and Ruskin H J 2008 *Physica* A **387** 4248–60
[20] Conlon T, Ruskin H J and Crane M 2007 *Physica* A **382** 565–76
[21] Sengupta A and Mitra P P 1999 *Phys. Rev.* E **60** 3389–92
[22] Strang G 1980 *Linear Algebra and its Applications* (New York: Academic Press)
[23] Pafka S and Kondor I 2002 *Eur. Phys. J.* B **27** 27780
[24] Pafka S and Kondor I 2002 *Physica* A **319** 487–94

A Media Information Analysis for Implementing Effective Countermeasure against Harmful Rumor

Mitsuyoshi Nagao[1], Kazuhiro Suto[2] and Azuma Ohuchi[3]

[1]Faculty of Business Administration and Information Science, Hokkaido Information University, 59-2 Nishi-Nopporo, Ebetsu, Hokkaido 069-8585, Japan

[2]Graduate School of Business Administration and Information Science, Hokkaido Information University, 59-2 Nishi-Nopporo, Ebetsu, Hokkaido 069-8585, Japan

[3]Faculty of Commercial Science, Hokkai School of Commerce, 6-6, Toyohita, Toyohira-ku, Sapporo, Hokkaido 062-8607, Japan

E-mail: nagao@do-johodai.ac.jp, k-suto@do-johodai.ac.jp and ohuchi@hokkai.ac.jp

Abstract. When large scale earthquake occurred, the word of "harmful rumor" came to be frequently heard. The harmful rumor means an economic damage which is caused by the action that people regard actually safe foods or areas as dangerous and then abort consumption or sightseeing. In the case of harmful rumor caused by earthquake, especially, tourism industry receives massive economic damage. Currently, harmful rumor which gives substantial economic damage have become serious social issue which must be solved. In this paper, we propose a countermeasure method for harmful rumor on the basis of media trend in order to implement speedy recovery from harmful rumor. Here, we investigate the amount and content of information which is transmitted to the general public by the media when an earthquake occurred. In addition, the media information in three earthquakes is treated as instance. Finally, we discuss an effective countermeasure method for dispeling harmful rumor through these analysis results.

1. Introduction

Several large earthquakes have occurred in Japan for the past few years, and they have given serious damage to various regions. For example, the 2007 Chuetsu Offshore Earthquake is included as the earthquake which is fresh in one's memory.

When large scale earthquake occurred, the word of "harmful rumor" came to be frequently heard. The harmful rumor means an economic damage which is caused by the action that people regard actually safe foods or areas as dangerous and then abort consumption or sightseeing [1]. In the case of earthquake, especially, the harmful rumor often arises in tourism industry. Currently, harmful rumor which gives substantial economic damage to neighboring tourism area which does not have physical damage around disaster-stricken area have become serious social issue which must be solved because harmful rumor gives bad influence to the revival of disaster-stricken area.

Despite this, the problem of harmful rumor is not supported by the laws or statutes of Japan government [2]. Moreover, academic studies on harmful rumor and its countermeasure are few. Therefore, the people who receives harmful rumor have to carry out the countermeasure through trial and error [3]. For this reason, the establishment of effective countermeasure for realizing early recovery from harmful rumor is required immediately.

It is obvious that the information transmitted to the general public by the media is included as one of major causes for the occurrence and expansion of harmful rumor. Immediately after the occurence of earthquake, the biased information such as the earthquake magnitude and serious damage condition in disaster-stricken areas is transmitted to the general public. The biased information is reported by various media such as newspaper and web all the time. The general public misunderstand that neighboring sightseeing areas which do not actually receive damage have physical damages as serious as the disaster-stricken areas by looking at or hearing the biased coverage.

It is likely that the effect of countermeasure of harmful rumor depends on the media condition because the occurrence and expansion of harmful rumor relates to the media information. Therefore, the countermeasure for harmful rumor must be performed in consideration of the condition of media information [4]. Namely, sufficient effect of countermeasure may not be obtained even if the countermeasure is conducted in the biased condition that the media transmits only devastating damage aspects. It seems reasonable to suppose that the countermeasure should be performed when the media condition became appropriate condition that the coverage which give bad influence against sightseeing blows over. However, the studies on the countermeasure for harmful rumor based on the media trend have been not performed.

In this paper, we propose a countermeasure method for harmful rumor on the basis of media trend in order to implement speedy recovery from harmful rumor. In our research, first, we investigate the amount and contents of information which is transmitted to the general public by the media when an earthquake occurred. Here, we employ newspaper and web news as investigation target. Newspaper has a feature which can transmit the information to the wide range of age, while web news has a feature that the circulation of information is speedy. We also analyze the difference among two medium which have different character. Additionally, we perform classification of the information reported by each media. In this classification, the information is classified on the basis of whether the article gives bad influence to sightseeing or not. The temporal alterations in these analyses is examined.

These analyses are applied to three earthquake disasters which have recently occurred in Japan. Their earthquakes are "2007 Noto Earthquake", "2007 Niigata Chuetsu Offshore Earthquake" and "2008 Iwate-Miyagi Nairiku Earthquake". They were the earthquakes which gave serious physical damage to cities near seismic center. On the other hand, in these earthquakes, the sightseeing areas which are located comparatively near their disaster-stricken areas and do not almost have physical damage received harmful rumor. We reveal the relationship among media information, physical damage and harmful rumor through the analysis of media information in these earthquakes.

Finally, we discuss an effective countermeasure method for dispeling harmful rumor through these analysis results.

2. Harmful Rumor

2.1. Outline of Harmful Rumor

When affair, trouble and disaster occurred, the fact is extensively reported by various media. The general public hears and sees the report, and then stop consumption or tourism because they regard the fundamentally safe product or area as dangerous. The word of "Harmful rumor" currently used indicates an economic damage which is caused by such behavior of the general public. Especially, inaccurate reports and exaggerated reports expands harmful rumor, and then the magnitude of damage of harmful rumor becomes huge. However, the report by media does not have all source of harmful rumor. Harmful rumor is also generated by the problem on the side of the receiver of information even if media transmitted accurate information. The trouble of Chinese-made poisoned frozen dumpling which occurred in Japan is included as such example.

A health hazard of food poisoning occurred on January 2008 in Chiba prefecture of Japan.

After investigation, the poison used for bug killer was detected from Chinese-made frozen meat dumpling which was eaten as a meal. Many patients complaining of food poisoning appeared on various region in Japan after the coverage of this food poisoning. However, there was no patients having a food poisoning caused by the poison which is used for the bug killer. Due to this coverage, not only Chinese-made other frozen food did receive harmful rumor, but also Japanese food maker treating frozen food, and the restaurant of Chinese food received economic damage.

In this affair, the report that a Chinese-made frozen meat dumpling is dangerous was transmitted to the general public. Ordinarily, the general public ought to understand that a meat dumpling made by a specific Chinese food maker is dangerous. However, the general public thought that all Chinese-made food and all meat dumpling which does not almost have the relationship is dangerous. Harmful rumor is generated by an extended interpretation by the receiver of information even if the media reports detailed information. It is actually impossible to prevent the occurrence of harmful rumor because it is difficult to completely remove the ambiguity from the information and coverage. Therefore, a risk management for harmful rumor is required.

2.2. Harmful Rumor by Earthquake

The same may be said of harmful rumor caused by earthquake. When earthquake happened, each media intensively perform the coverage about the area which received heavy damage. Therefore, it is likely that the general public misunderstands the scale of damage in the area. Namely, the scale of damage is recognized greater than actual situation. Moreover, the general public can guess the rough location of area which received damage. However, it is difficult to accurately guess the range of damage area from contents of coverage. Thus, the general public misunderstands that the damage area is wider than actual damage region. As these results, the general public misunderstand that earthquake gave immense damage to wide range.

Harmful rumor by earthquake gives immense economic damage to local economy. Enormous cost is required for the revival of the area which received physical damage. However, in the area which was saved from physical damage, tourism industry receives considerable economic damage. The income from tourism industry is cut off. As these results, the local economy is confused. The vicious circle is generated that the recovery of disaster-stricken area becomes sluggish and harmful rumor extend over a long period of time. However, the study on harmful rumor which is generated by earthquake has been few. Consequently, the countermeasure of harmful rumor has been performed through trial and error. Currently, the establishment of effective countermeasure for harmful rumor is demanded.

3. Countermeasure for Harmful Rumor on the basis of Media Trend

It is important to perform the countermeasure which can realize early recovery from harmful rumor because it is difficult to prevent harmful rumor from being generated. Therefore, it is required to perform public relations activities for attracting tourists by providing information to consumers.

The trend of the information transmitted by media should be checked when such countermeasure is performed. Tourists are easily influenced by the information transmitted by media. Namely, it is obvious that tourists restrain visiting to neighboring area around disaster-stricken area while the media transmits biased coverage of disaster related information such as damage condition. Even if the countermeasure for harmful rumor was performed in the period when the biased coverage is transmitted, enough effect can not be obtained. Moreover, it is likely that the general public has doubt about the countermeasure activity and the countermeasure has the opposite effect. As time advances, the number of disaster related information decreases, and the contents of coverage change from the situation of damage to the situation of restoration.

It is possible to get enough effect without the obstruction by disaster related information, if the countermeasure for harmful rumor is put into practice during such period.

When the condition of media changed for the better, the enforcement of countermeasure for harmful rumor which utilize the media is effective. The appropriate countermeasure using the media in the period which biased condition in media blows over can promote tourism activity of consumer because consumer is easily influenced by the media.

So far, various countermeasures such as public relations activity on the street have been performed in order to wipe out harmful rumor. However, there were many countermeasures which had no effect because they were executed while the media extensively transmitted the biased information about the situation of disaster.

Therefore, firstly, it is necessary to grasp the trend of disaster related information transmitted by the media, and then clarify the period which can sufficiently obtain the effect of countermeasure. In addition, the discussion about the contents of effective countermeasure activity in such period is needed.

In this research, we investigate and analyze the coverage of the media in three earthquakes which recently occurred in Japan. In addition, we investigate the relationship between damage situation and media coverage, and grasp the trend of media. The contents and start period of countermeasure is discussed on the basis of the analysis results. From our achievement, the countermeasure based on the media trend can be implemented, and harmful rumor in the tourism area around disaster-stricken area can be effectively wiped out in comparison with the case that the countermeasure through trial and error is performed.

4. Target Earthquakes for Media Information Analysis

4.1. 2007 Noto Earthquake

The 2007 Noto Earthquake is a massive earthquake which occurred in 25th March, 2007. The Noto is a peninsula which is located on the side of Sea of Japan (See Figure 1). Almost whole area of peninsula belongs to the Ishikawa prefecture. The seismic center is the Sea of Japan near Noto peninsula. On the Richter scale, the magnitude of the earthquake was estimated at 6.9. The earthquake shook the city of Nanao and the town of Anamizu on the Ishikawa prefecture with a seismic intensity of 6 upper in Japan's shindo scale. The earthquake brought down great damage to many cities and towns in the Ishikawa prefecture and the Toyama prefecture which is neighboring prefecture of Ishikawa prefecture. At least, 1 dead and 279 injured in the Ishikawa and the Toyama prefectures have been reported as human sufferings. As physical damage, 649 completely destroyed houses and 26,614 partly destroyed houses have been reported. The lifeline utilities, e.g., electricity, water, gas supplies and so on, were destroyed. In addition, all railways in this area stopped. In this earthquake, harmful rumor happened in many tourist resorts in Ishikawa prefecture. For instance, in the Wakura Hot-spring resort in Ishikawa prefecture, there were 66,413 lodging cancelations in hotels during a month from the earthquake occurrence. The Wajima City of Ishikawa prefecture had 15,526 lodging cancelations in hotels. Moreover, the sales related to tourism decreased 20% in Kaga Hot-spring resorts in Ishikawa prefecture which received almost no physical damage and is far from the seismic center.

4.2. 2007 Niitaga Chuetsu Offshore Earthquake

The 2007 Niigata Chuetsu Offshore Earthquake (described as the 2007 Niitaga Earthquake) occurred in 16th July, 2007. Niigata prefecture is located on the side of the Sea of Japan (See Figure 1). Chuetu is the center area in three areas of Niigata prefecture. The seismic center is the Sea of Japan off the coast of Chuetsu. The magnitude of the earthquake was estimated at 6.8. The city of Kashiwazaki and Nagaoka and the villages of Kariwa registered the highest seismic intensity of 6 upper on Japan's shindo scale. Fifteen deaths and at least 2,315 injuries have been reported. In addition, 1,319 buildings were completely destroyed and 40,280 buildings

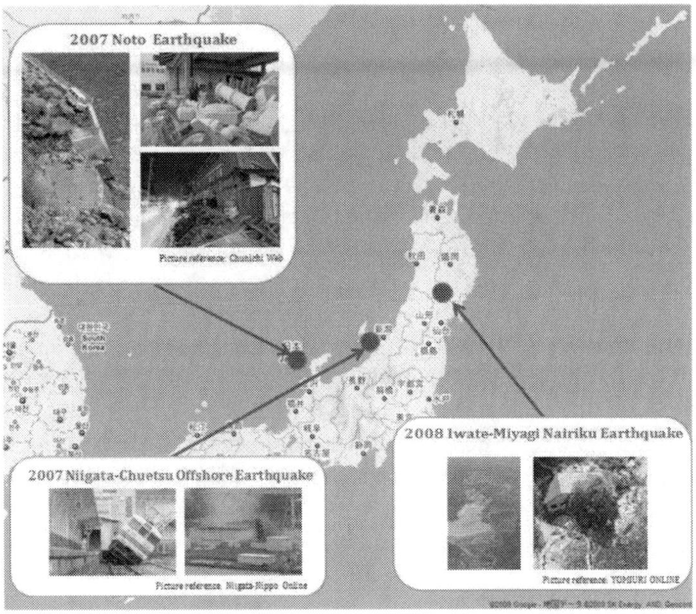

Figure 1. Position where earthquakes occurred

were partly destroyed. The lifeline utilities and various transport facilities have been broken down. Furthermore, a remarkable damage happened in this earthquake. It is Kashiwazaki-Kariya Nuclear Power Plant incidents. A fire broke out in an electrical transformer at the Kashiwazaki-Kariwa Nuclear Power Plant. A leak of radioactive gases happened by this. Many tourist resorts in the Niigata prefecture were hard hit by harmful rumor due to fears relating to the plant accident. For example, it was reported that the number of hotel guests in whole of hot spring resorts in the Niigata prefecture decreased 40%.

4.3. 2008 Iwate-Miyagi Nairiku Earthquake

The 2008 Iwate-Miyagi Nairiku Earthquake (described as the 2008 Iwate Earthquake) occurred in 14th June, 2008. Iwate and Miyagi is one of prefectures in Japan, the seismic center was border between these prefectures. The magnitude of the earthquake was estimated at 7.2. The city of Oshu and Kurihara registered the highest seismic intensity of 6 upper on Japan's shindo scale. Thirteen deaths and 448 injured have been reported as human sufferings. In addition, 23 buildings were completely destroyed and 1,155 buildings were partly destroyed. In this earthquake, massive landslide occurred around the seismic center. Some rivers were dammed by mudslide, and then some landslide dams were generated. There was a fear that secondary disaster happens by the collapse of landslide dams. On the other hand, urban area did not almost receive physical damage because the seismic center was mountain region.

Many reservations for tourism were canceled in many cities and tourist resorts in the prefectures around the seismic center although there was almost no damage in urban area. As one misfortune followed another, a big earthquake which was 6 upper on Japan's shindo scale happened after one month. By this earthquake, tourism industry in the region around seismic center received extensive damage by harmful rumor.

5. Media Information Analysis

5.1. Analysis Method

We collected the information related to each earthquake in two kinds of media. One is newspaper and another is online news website (described as web news) which is managed by newspaper publishing company. In each earthquake, we collected the articles related to earthquake from five newspapers. They are four national newspapers which are Asahi Shinbun, Yomiuri Shinbun, Nihon Keizai Shinbun and Mainichi Shinbun, and a local newspaper which is Hokkaido Shinbun. In the analysis of web news, we used two websites for 2007 Noto Earthquake, three websites for 2007 Niigata Earthquake, and five websites for the 2008 Iwate Earthquake, respectively. We collected the articles from CHUNICHI Web (http://www.chunichi.co.jp/) and YOMIURI Online (http://www.yomiuri.co.jp/) in the 2007 Noto Earthquake. In the 2007 Niigata Earthquake, we collected the articles related to earthquake from MSN Sankei News (http://sankei.jp.msn.com/), Niigata-Nippo Online (http://www.niigata-nippo.co.jp/) and YOMIURI Online. Finally, in the 2008 Iwate Earthquake, the articles were collected from MSN Sankei, Mainichi JP, KAHOKU ONLINE NETWORK (http://www.kahoku.co.jp/), CHUNICHI Online and YOMIURI Online. YOMIURI Online, MSN Sankei News, Mainichi JP are online news website managed by national newspaper publishing company in Japan, and CHUNICHI Web, Niigata-Nippo Online and KAHOKU ONLINE NETWORK are online news website for local newspaper publishing company. In this collection, we treated the articles including the name of the earthquake as the disaster related information. The period to collect the articles was 50 days from the occurrence of each earthquake.

We investigated the information amount in newspaper and web news. Here, we used the area of articles as the information amount in newspaper. And, we used the number of articles as the information amount in web news. Moreover, we analyzed the contents of articles collected from both media. In this contents analysis, we classified the articles into three categories on the basis of whether it gives avoidance emotion to the general public against sightseeing or not. The articles giving avoidance emotion to the public are classified into "-" category. For example, the articles which the situation of damage and the number of deaths are reported are included in this category. The articles which alleviate avoidance emotion are classified into "+" category. This category include the articles about the public relations activity and the advance of revival. Finally, the articles which do not relate to avoidance emotion are classified into "0" category. The articles which report the movement of economy and politics related to the occurrence of earthquake are included in this category.

5.2. Analysis Result

Figure 2 indicates the transition of information amount in newspaper for three earthquakes. In this figure, the x-axis means the number of days after the occurrence of earthquake, and the y-axis means the total size of articles published with five news papers. As shown in this figure, the information amount in the 2007 Noto Earthquake remarkably decreased during 3 days from the occurrence of earthquake, and the amount did not increase after this. In the case of 2007 Noto Earthquake, property damage happened over wide range, but human suffering was few. It is likely that the amount of coverage decreased because of the lack of fresh information. In the case of 2007 Niigata Earthquake, the amount of information increased after one week and after one month of the earthquake occurrence, and the amount did not increase after this. The increase after 48 days is exception because the reason of this increase was caused by tourism public relations advertisement published by government. In this earthquake, the prolonged coverage was performed. We can be fairly certain that the prolonged coverage was performed because this earthquake had the characteristic damage, e.g., the accident in nuclear power plant and the stoppage of car manufacture which give also bad influence to other area. Finally, in the 2008 Iwate Earthquake, the information amount remarkably increased after one week of the

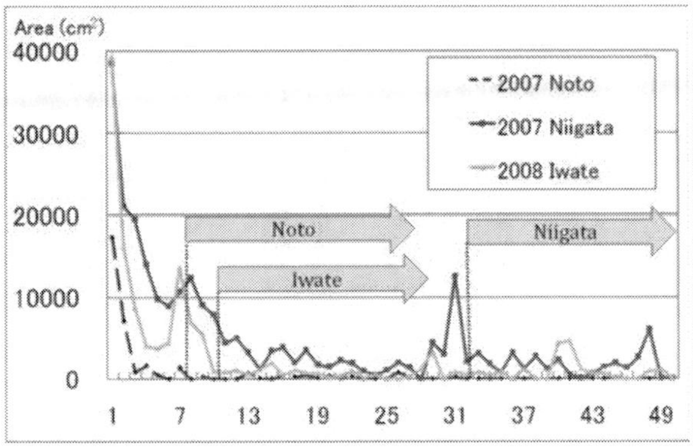

Figure 2. Transition of information amount in newspaper

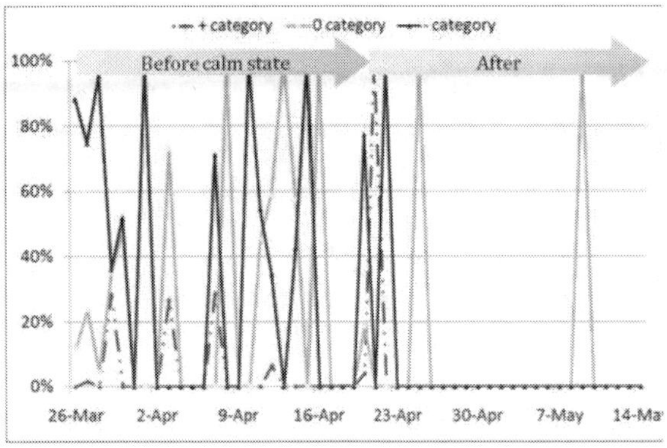

Figure 3. Result of contents analysis in newspaper for 2007 Noto Earthquake

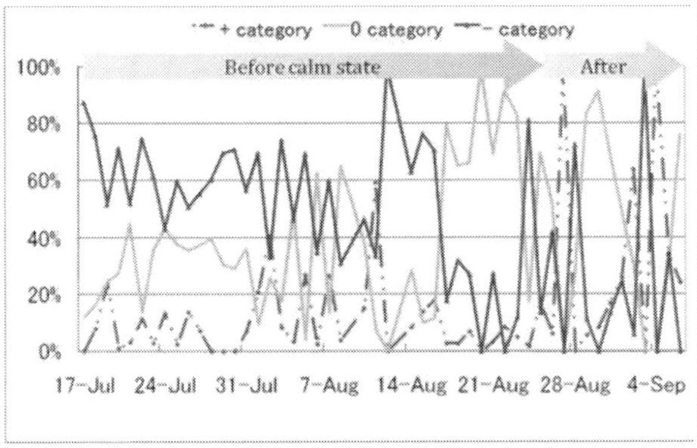

Figure 4. Result of contents analysis in newspaper for 2007 Niigata Earthquake

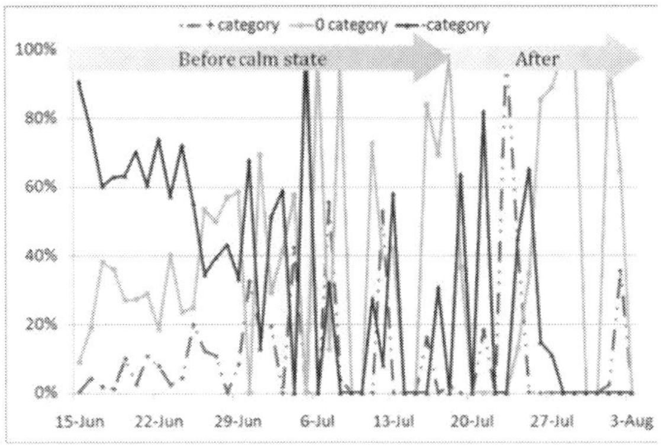

Figure 5. Result of contents analysis in newspaper for 2008 Iwate Earthquake

occurrence. The damage in this earthquake concentrated to narrow area of mountain-ringed region, and the neighboring urban area did not almost receive physical damage. However, human suffering was higher than the 2007 Noto Earthquake, and the rescue operation of the sufferers had hard going. The summary of situation of damage was reported after one week of the occurrence. As compared with three results, in the case where the coverage extends over a long period of time, it was found that the summary of earthquake disaster was reported. In addition, it was revealed that the continuity of coverage depends on the scale and feature of damage in earthquake.

We analyzed the appropriate period for performing the countermeasure for harmful rumor from the result of information amount. We regarded as calm state of media the situation that the information amount is adequately low value and does not rapidly increase again. From the results, we judged that the appropriate period was one week after the occurrence in the 2007 Noto Earthquake, the period in the 2007 Niigata Earthquake was after one month, and the period in 2008 Iwate Earthquake was after 10 days, respectively.

Results of contents analysis in newspaper for three earthquakes are represented in Figure 3, 4 and 5. These figures show the ratio of the information amount in each category. In Figure 3, the situation of contents of newspapers in the 2007 Noto Earthquake can be divided into three periods on the basis of the ratio on each category. First period is a time between the day of occurence and one week after. Second period is a time between one week and about three weeks

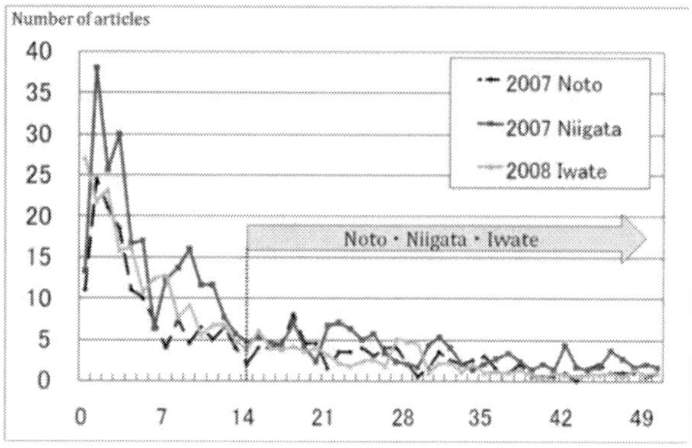

Figure 6. Transition of information amount in web news

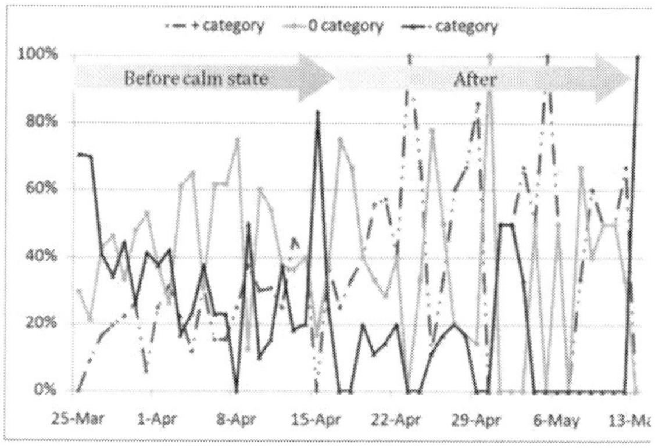

Figure 7. Result of contents analysis in web news for 2007 Noto Earthquake

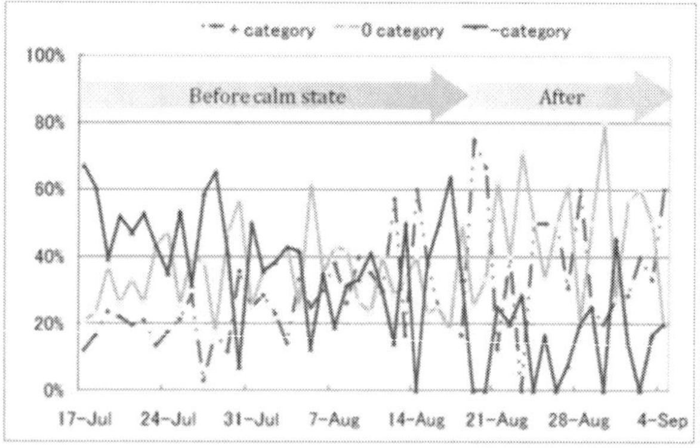

Figure 8. Result of contents analysis in web news for 2007 Niigata Earthquake

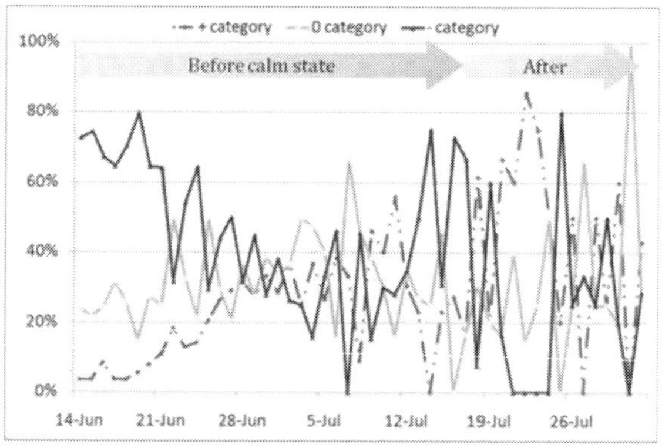

Figure 9. Result of contents analysis in web news for 2008 Iwate Earthquake

after from the occurrence. Final period is after three weeks from the occurrence. In the first period, the ratio of "-" category was continuously high, and then the ratios except "+" category had large oscillation in second period. In final period, the ratios of all categories had large oscillation. As shown in Figure 4, the result of the contents analysis could be also divided into three periods in the 2007 Niigata Earthquake. In the first period, i.e., until after four weeks of the earthquake occurence, the ratio of "-" category was high, and then the ratio of "0" category came high in second period, i.e., during two weeks after the end of first period. After this, the ratio of each category had large oscillation. In the 2008 Iwate Earthquake, the result can be divided into the five periods on the basis of the ratio on each category. The situation that the ratio of "-" was large continued during one week from the occurrence of earthquake. After this, the ratios except "+" had oscillation until after three weeks of the occurrence. The three ratios had oscillation until after four weeks, and then the ratio of "-" increased again. Finally, three ratios had oscillation again from after four weeks half of the occurrence. From these results, we judge calm period of the media information from the contents point of view. Here, we regarded as the calm period of contents in the disaster related information the situations that the ratio of "-" category is low and the ratio of "+" category emerges. In addition, this calm period is judged with the consideration of continuity of the ratio in each category. As this result, it was revealed that it was three weeks half after in the 2007 Noto Earthquake, six weeks later in the 2007

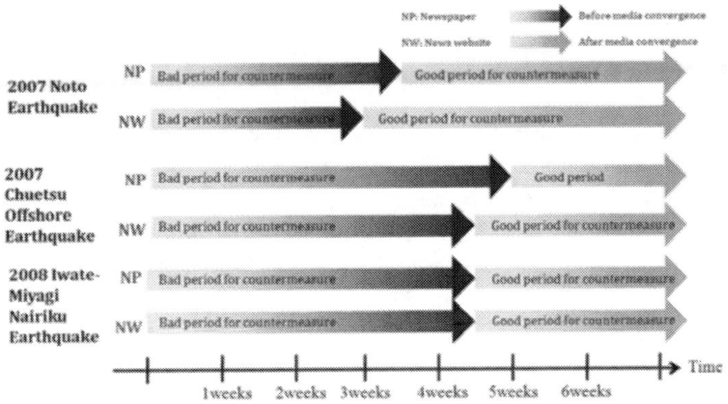

Figure 10. Summarized result of appropriate period for countermeasure

Niigata Earthquake, and four weeks half after in the 2008 Iwate Earthquake, respectively. As the appropriate period for countermeasure of harmful rumor, it is likely that the calm situation of media information needs to be confirmed in both aspects of amount and contents. Therefore, in the analysis results of newspaper, the appropriate period for countermeasure of harmful rumor corresponded with the calm period of contents.

Figure 6 shows the transition of information amount in web news. In this figure, the average number of articles per one website is represented. Although three earthquakes happened in the morning, the maximum value of information amount emerged on the day of occurrence in the 2008 Iwate Earthquake, and emerged on the next day of occurrence in other earthquakes. This difference depends on the time required to confirm whole of damage in earthquake. The 2007 Noto Earthquake and 2007 Niigata Earthquake caused damage over large areas. Therefore, the maximum value emerged on the next day of the occurrence because it took long time to investigate a whole of damage situation. On the other hand, in the 2008 Iwate Earthquake, the confirmation of damage situation was smoothly performed because damage area concentrated to narrow area of the mountain range. Thus, the maximum value was confirmed on the day of the occurence. Although the trend of information amount in three earthquakes was different until two weeks after the occurence of earthquake, the slow decrease trend could be found in all after this. This means that disaster related information on web news was calm by two weeks. Judging from information amount, the calm period for media information was after two weeks of the occurrence.

Figure 7, 8 and 9 indicates the result of contents analysis in web news for three earthquakes. When the calm period of contents in media information is regarded as the situations that the ratio of "-" category is low and the ratio of "+" category emerges, it was found that the period was after three weeks of the occurrence in the 2007 Noto Earthquake. For remaining two earthquakes, it was four weeks later. The appropriate timing for the countermeasure of harmful rumor corresponded with the calm period of content as in the result of newspaper, because the biased condition in both of amount and contents must be blown over.

Figure 10 shows the summarized result of appropriate period for countermeasure of harmful rumor in three earthquake.

6. Discussion

We discuss the countermeasure method using each media for harmful rumor on the basis of the analysis results.

The newspaper has the characteristics that the article having large area attracts someone's attention, and it is difficult for article to be published if it does not have high topicality. Therefore, as the countermeasure using newspaper, the insertion of public relations advertisement which has enough wide area to attract a great deal of attention in the calm period of media which enough effect of the countermeasure is expected is one of effective methods. The advertising activity can be imprinted to the general public without being obstructed by news which give bad influence to sightseeing. Such countermeasure can remove negative impression of visiting the tourism spot around disaster-stricken area and can give positive impression for tourism to the general public. Moreover, the effect to the people of all ages is expected because newspaper is read by various people.

In web news, it is likely that events and affairs are published as the articles for news website even if they do not have high topicality. In addition, they are reported promptly on the website. Therefore, the enforcement of lively public relations activity after calm period of media information is effective method as the countermeasure using web news. By performing such countermeasure, the effect of public relations activity can be obtained firstly, and then the ripple effect by the publication of the article to news website can be expected.

Moreover, the harmful rumor by earthquake is concerned to the factor of hesitating to go sightseeing near disaster-stricken area even if the tourism area is safe. Therefore, it is required to clarify to the general public that sightseeing to the tourism area contributes to smooth rehabilitation of the disaster-stricken area.

7. Conclusion

In this paper, we proposed a countermeasure method for harmful rumor on the basis of the trend of disaster related information in order to realize early recovery from harmful rumor. To implement our countermeasure method, we analyzed the disaster related information of newspaper and web news in three earthquakes, from the information amount and content points of view. From the analysis results of disaster related information on newspaper, the calm period of media which is suitable for starting the countermeasure for harmful rumor could be confirmed after 3 weeks from the occurrence of earthquake in early case. The calm period was after 6 weeks in late case. On the other hand, from the analysis results of web news, it was revealed that the calm period of media was after 3 weeks in early case and it was 5 weeks in late case. The analysis results indicated that the result of contents analysis was more important to the starting period of countermeasure. However, the content analysis requires higher cost in comparison with the amount analysis. Therefore, in order to realize a low cost and real time analysis for the countermeasure of harmful rumor, the relationship between amount and content must be clarified. In addition, the implementation of the prediction of media trend based on the scale and feature of physical damage is required in order to realize the effective countermeasure for harmful rumor.

In the contents analysis, the disaster related information was classified by the criteria which we decided in advance. Namely, it was not completely objective classification. Therefore, we plan to implement completely objective analysis by employing text-mining approach for contents analysis.

Moreover, in this research, we discussed the countermeasure method of harmful rumor using the media. We judged that publishing public relations advertisement having a lot of space is effective as the countermeasure method using newspaper because newspaper has limitation for publishing space. As the countermeasure using web news, the enforcement of many public relations activities is effective because web news posts report about an event even if it is not impactful. We plan to investigate more detailed contents and effect of public relations activity by performing the hearing investigation in the area which received harmful rumor generated by earthquake. These are our future work.

References

[1] N. Sekiya, "Social Psychology of "Image Contamination" - The Fact of "Image Contamination" and its Mechanism", Journal of Japan Society for Disaster Information Studies, No.1, pp.78-89.

[2] N. Ishii, "Governor Asks METI to Help Dispel Harmful Rumors after Earthquake Damage", Japan Leading Nuclear information Source, Atoms In Japan, http://www.jaif.or.jp/ english/aij/member/2007/2007-07-24a.pdf, 2007

[3] M. Nagao, H. Kishino and A. Ohuchi, "Approach of Echigo-yuzawa on the Mid-Niigata Earthquake Rumor Damage -Development of General Rules from the Case in Echigo-yuzawa-", Journal of Society for Tourism Informatics, pp.30-40, 2006.

[4] M. Nagao, K. Suto, A. Yamashita, Y. Matsumura and A. Ohuchi, "Media Information Analysis for Countermeasure against Damage Caused by Rumor on Tourist Spot", Proceedings of Information Processing Society of Japan Hokkaido Symposium 2007.

Cascade dynamics on clustered network

Y Ikeda[1], **T Hasegawa**[2], **and K Nemoto**[1]

[1] Department of Quantum and Condensed Matter Physics, Graduate School of Science, Hokkaido University, Kita 10-jo Nisi 8-tyome, Sapporo, JAPAN

[2] Graduate School of Information Science and Technology, The University of Tokyo, 7-3-1 Hongo, Bunkyo-ku, Tokyo, JAPAN

E-mail: `hasegawa@stat.t.u-tokyo.ac.jp`

Abstract. We investigate information cascades on a clustered network model generated from a projection of bipartite graphs by numerical simulations and compare it with that on a non-clustered network having the same degree distribution. Our result indicates that global cascades occur more easily on clustered networks than on non-clustered ones. Furthermore, we extend a recursive method by Gleeson and Cahalane to estimate the order parameter analytically and find that the result gives an excellent approximation to the observed transition point.

1. Introduction

In many complex networked systems [1–3], small initialized shocks are often propagated through a whole system. For example, the spread of rumors and the diffusion of norms and innovations are observed in social systems. These phenomena are called *information cascades*, during which individuals in a population exhibit collective behavior [4–17]. To explain the mechanism causing information cascades, Watts [4] proposed a network model (the Watts model) in which decisions of interacting agents are determined by the actions of their neighbors according to a simple threshold rule:

(1) Consider a network consisting of nodes and edges, which correspond to individuals and relationships between them, respectively. The state of each node is either active or inactive.

(2) As an initial state, we set a randomly-chosen node active. The node is called a *seed*.

(3) At each time step, each inactive node updates its state active if at least a fraction ϕ of its neighbors are active, otherwise it remains inactive. Once being active, a node cannot turn back to the inactive state (the *permanently active property*).

(4) Step 3 is iterated until no node changes its state. Then cascade size ρ is evaluated as the final fraction of active nodes.

The dynamics of cascades depends on the structure of underlying network [4–8]. For random graphs it is well known that very large cascades, so called *global cascades* having non-vanishing ρ, are triggered by small initialized shocks for small ϕ and disappear as a discontinuous phase transition in ρ with increasing ϕ [4, 5].

The dynamics of cascades has been well analyzed on locally tree-like networks, whereas many social networks have high clustering coefficient. Among such high-clustering networks, there are many real social networks represented as bipartite graphs, *e.g.*, networks of CEOs, boards of directors, and collaborations of scientists [2]. A bipartite graph is defined as a network in which

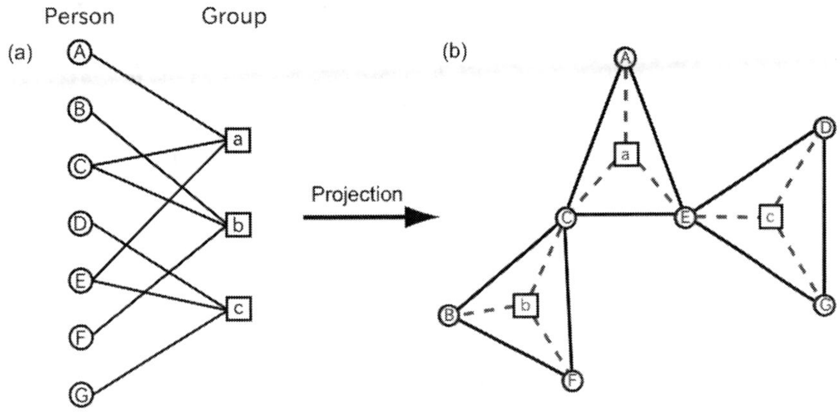

Figure 1. (a) Bipartite graph consisting of Person-nodes (circle) and Group-nodes (square), and (b) its projection onto Person-nodes. The projected network consists of only triangles under the assumption that all Group-nodes have degree 3.

nodes are divided into two sets in the way that no edge connects two nodes in the same set. In this paper, we analyze the Watts model on a clustered network generated from a bipartite graph to find the effect of clustering on cascade dynamics. Our numerical and analytical results demonstrate that global cascades on a clustered network occur in wider range of ϕ compared with those on a non-clustered one. Moreover our result shows that global cascades which are triggered by the effect of clustering structure are rare events, indicating that information cascades in our society are often hard to anticipate.

2. One-mode projected network

In this section, we introduce *projected network* and *reconstructed network*, as a clustered network and non-clustered one, respectively. Our clustered network model is generated by a projection of a bipartite graph [18]. Firstly, we consider a bipartite graph which consists of *Person-* and *Group-*nodes (Fig.1-(a)). Person-nodes follow a Poisson distribution $P_{\mathrm{Person}}(k) \propto (z/2)^k \mathrm{e}^{-z/2}/k!$ with mean $z/2$, and all Group-nodes have the same three degree $P_{\mathrm{Group}}(k) = \delta_{k,3}$. Now the bipartite graph can be projected onto Person-nodes only (Fig.1-(b)). This resulting graph has a Poisson degree distribution $P(k) \propto z^k \mathrm{e}^{-z}/k!$ $(k = 2, 4, 6, \cdots)$, and simultaneously has a high density of triangles, the clustering coefficient $C = 1/(z+1)$ [18]. We call this graph the projected network. Next, keeping the degree of each node, we reconnect edges at random. This operation produces a non-clustered network with the same degree distributions as the original one, which we call the reconstructed network.

We trigger information cascades on these two networks, and compare the results to investigate the effects of clustering structure on information cascades. In the next section, we propose an approximation to analyze the information dynamics on the projected network.

3. Random cactus approximation

Gleeson and Cahalane [5] proposed a general analytical method by using recursion relations to analyze the cascade dynamics for locally tree-like networks with arbitrary degree distributions. We extend their method to estimate the order parameter ρ on the projected network analytically.

Cascade size ρ is equal to the probability that a randomly-chosen node is active in a final state. To estimate this probability, we choose an arbitrary node as the root, and approximate the projected network by a random cactus tree with $P(k)$, starting from the root (see Fig.2). We

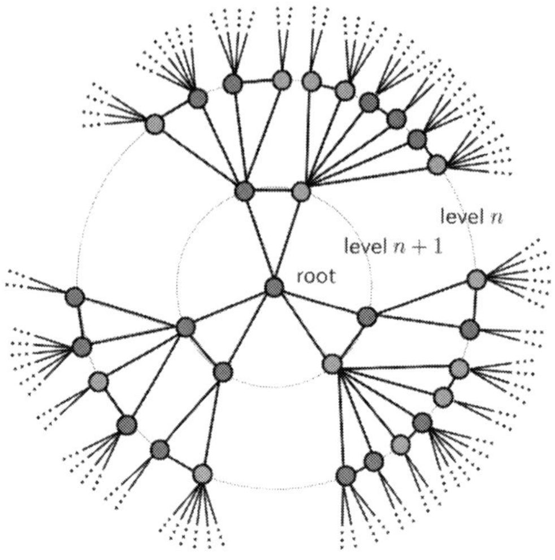

Figure 2. Random cactus approximation.

label the levels n of the nodes from the bottom of the tree ($n = 0$), with the root at an infinitely high level ($n = \infty$). We estimate the propagation of a cascade starting from a single seed ($\rho_0 = 1/N$) at the bottom by the following update equations. Firstly, we define the probabilities $q_n^{(i)} (i = 1, 2, 3, 4)$ that a node pair at level n is in the state i, given that the parent node at level $n+1$ is inactive. Here the pair-states $i = 1, 2, 3$, and 4 correspond to inactive-inactive, active-inactive, inactive-active, and active-active node-pairs, respectively. The probabilities of the pair-states at level n are given by the probabilities $\tilde{q}_n$ and $\tilde{q}'_n$, where $\tilde{q}_n$ ($\tilde{q}'_n$) is the probability that an inactive node at the n-th level changes its state on the condition that the other of the pair being inactive (active). The update equations for $\tilde{q}_n$ and $\tilde{q}'_n$ are given as

$$\begin{aligned} \tilde{q}_{n+1} &= \rho_0 + (1-\rho_0) \sum_{k=0}^{\infty} \frac{(k+1)p_{k+1}}{z} \times \\ & \sum_{k_1,k_2,k_3,k_4} \delta_{k-k_1-k_2-k_3-k_4} \frac{k!}{k_1!k_2!k_3!k_4!} {q_n^{(1)}}^{k_1} {q_n^{(2)}}^{k_2} {q_n^{(3)}}^{k_3} {q_n^{(4)}}^{k_4} F(k_2+k_3+2k_4, 2k+2), \end{aligned} \tag{1}$$

$$\begin{aligned} \tilde{q}'_{n+1} &= \rho_0 + (1-\rho_0) \sum_{k=0}^{\infty} \frac{(k+1)p_{k+1}}{z} \times \\ & \sum_{k_1,k_2,k_3,k_4} \delta_{k-k_1-k_2-k_3-k_4} \frac{k!}{k_1!k_2!k_3!k_4!} {q_n^{(1)}}^{k_1} {q_n^{(2)}}^{k_2} {q_n^{(3)}}^{k_3} {q_n^{(4)}}^{k_4} F(k_2+k_3+2k_4+1, 2k+2). \end{aligned} \tag{2}$$

Here the response function $F(m, k)$ is

$$F(m,k) = \begin{cases} 0 & \text{if } m < \phi k \\ 1 & \text{if } m \geq \phi k, \end{cases} \tag{3}$$

where m is the number of active nodes in k nearest neighbors. Then, we obtain probabilities $q_{n+1}^{(i)}$ at the $(n+1)$-th level as

$$q_{n+1}^{(1)} = (1 - \tilde{q}_{n+1})^2, \tag{4}$$

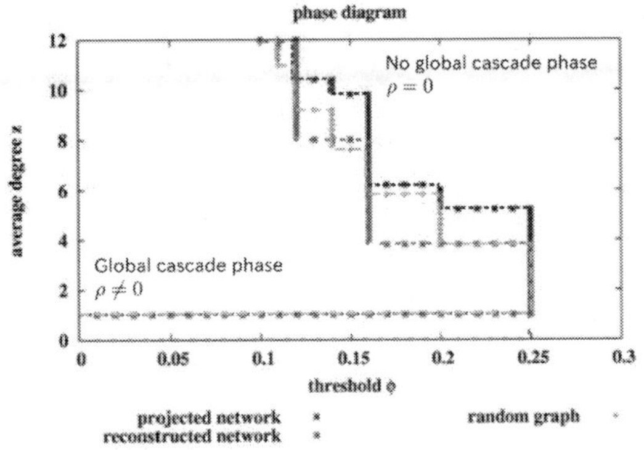

Figure 3. Phase diagram of $\phi - z$ plane. The symbols outline the region in which global cascades occur in simulation. The dashed line represents boundaries of cascade phase. The random graph shown here has a Poisson degree distribution $P(k) \propto 6^k e^{-6}/k!$ $(k = 1, 2, 3, \cdots)$.

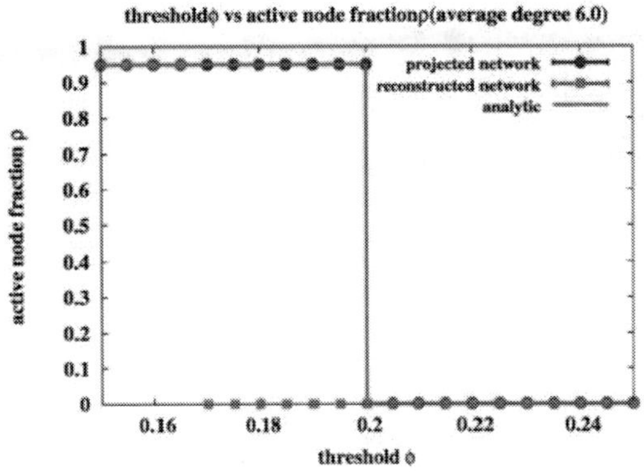

Figure 4. Active node fraction ρ as a function ϕ obtained numerically on the projected network (closed circles) and on the reconstructed network (closed squares) at $z = 6.0$. Our analytical result is also shown (solid line).

$$q^{(2)}_{n+1} = q^{(3)}_{n+1} = \tilde{q}_{n+1}(1 - \tilde{q}'_{n+1}), \tag{5}$$

$$q^{(4)}_{n+1} = \tilde{q}^2_{n+1} + 2\tilde{q}_{n+1}(\tilde{q}'_{n+1} - \tilde{q}_{n+1}). \tag{6}$$

We iterate this level-by-level update to obtain $q^{(i)}_\infty$ which are the fixed points of the above recursion relations. Then, the probability of the root node being active, i.e., the cascade size ρ, is given as

$$\rho = \rho_0 + (1 - \rho_0) \sum_{k=1}^{\infty} p_k \times \sum_{k_1,k_2,k_3,k_4} \delta_{k-k_1-k_2-k_3-k_4} \frac{k!}{k_1!k_2!k_3!k_4!} {q^{(1)}_\infty}^{k_1} {q^{(2)}_\infty}^{k_2} {q^{(3)}_\infty}^{k_3} {q^{(4)}_\infty}^{k_4} F(k_2 + k_3 + 2k_4, 2k). \tag{7}$$

4. Numerical simulations

In this section, we investigate numerically the cascade dynamics on both the projected network and the reconstructed network by Monte-Carlo simulation. We pick up the largest cascade in 10^2 cascade trials on each network sample with 10^6 nodes, and average those sizes over 10^2 network samples. The global cascade phase is given as the region in which the averaged cascade size is proportional to the network size.

Figure 3 expresses the cascade condition of the two network models graphically as a boundary in the phase diagram of threshold ϕ and average degree z. Clearly, the global cascade phase of the projected networks is wider than that of the reconstructed networks, indicating that cluster structure enhances information cascades.

Figure 4 shows our numerical and analytical results for active node fraction ρ as a function ϕ. One finds that the random cactus approximation gives an excellent approximation to the observed transition point, and thus supports our numerical results. Both results show that global cascades appear as a discontinuous phase transition in ρ with decreasing ϕ on both networks.

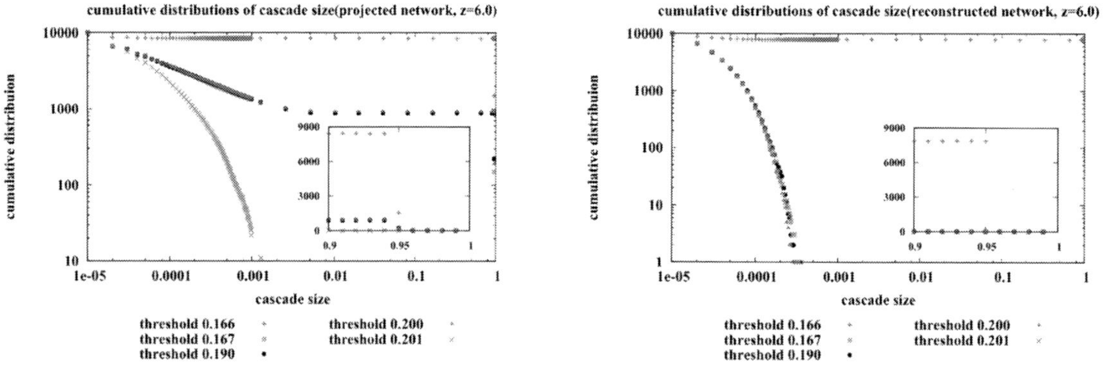

Figure 5. Cumulative distributions of cascade sizes on (a) the projected network and (b) the reconstructed network. For $z = 6.0$, we estimate $\phi_1 = 0.167$ and $\phi_2 = 0.2$, respectively (see text).

Finally, we compare the distributions of cascade sizes on both networks (Fig.5). From the distribution of cascade sizes, we find that the Watts model on the projected network takes the following three regimes:

- For $\phi > \phi_2$, where $\phi_2 = \phi_2(z)$ is the critical threshold of the projected networks, no global cascades occur on both the projected network and the reconstructed network.
- For $\phi_1 < \phi < \phi_2$, where $\phi_1 = \phi_1(z)$ is the critical threshold of the reconstructed networks, global cascades occur rarely (1 time per 10 trials) on the projected network while no global cascades occur on the reconstructed network. This means that global cascades in this extended regime are triggered by the effect of the cluster structure.
- For $\phi < \phi_1$, global cascades occur frequently (9 times per 10 trials) on both networks.

In [4], the cascade size distributions on random graphs show us that the information cascades occurs rarely at the critical threshold, implying a more extreme kind of instability that is harder to anticipate. Interestingly, our numerical result reveals that the cluster structure induces a bimodal distribution of cascade sizes, i.e., rarely occurring global cascades, in a wide range of ϕ. This result may indicate that global information propagations in our human society, where *two of your friends are likely to be friends of each other*, will be harder to predict.

5. Summary

We studied an information cascade model proposed by Watts [4] on a clustered network to find the effect of clustering on cascade dynamics. We extended a recursive method by Gleeson and Cahalane [5] to estimate the order parameter ρ analytically. Both analytical and numerical results showed that global cascades occur more easily on a clustered network than on a non-clustered one. Moreover, the clustered network has a nontrivial bimodal distribution of cascade sizes which indicates global cascades occur rarely in a wide range of ϕ.

As indicated in [6], recursive method for cascade propagations can be applied to other dynamics, *e.g.*, k-core percolations, by modifying the response function $F(m, k)$. It is very interesting to extend our method in this paper for general propagating dynamics.

Acknowledgments

The present work is supported by 21st Century COE program "Topological Science and Technology".

References

[1] Albert R and Barabási A -L 2002 *Rev. Mod. Phys.* **74** 47.
[2] Newman M E J 2003 *SIAM Review* **45** 167.
[3] Boccaletti S, Latora V, Moreno Y, Chavez M, and Hwang D -U 2006 *Phys. Rep.* **424**, 175.
[4] Watts D J 2002 *PNAS* **99** 5766.
[5] Gleeson J P and Cahalane D J 2007 *Phys. Rev.* E **75** 056103.
[6] Gleeson J P 2008 *Phys. Rev.* E **77** 046117.
[7] Dodds P S and Payne J L 2009 *Phys. Rev.* E **79** 066115.
[8] Payne J L, Dodds P S, and Eppstein M J 2009 *Phys. Rev.* E **80** 026125.
[9] Kempe D, Kleinberg J and Tardos E 2003 Proceedings of the ninth ACM SIGKDD international conference on Knowledge discovery and data mining 137.
[10] Dodds P S and Watts D J 2004 *Phys. Rev. Lett.* **92** 218701.
[11] Duan W Q, Chen Z, Liu Z, and Jin W 2005 *Phys. Rev.* E **72** 026133.
[12] Watts D J and Dodds P S 2007 *J. Consum. Res.* **34** 441.
[13] Centola D, Eguíluz V M, and Macy M W 2007 *Physica A* **374** 449.
[14] Galstyan A and Cohen P 2007 *Phys. Rev.* E **75** 036109.
[15] Klimek P, Lambiotte R, and Thurner S 2008 *Europhys. Lett.* **82** 28008.
[16] de Kerchove C, Krings G, Lambiotte R, Van Dooren P, and Blondel V D 2009 *Phys. Rev.* E **79** 016114.
[17] Galstyan A, Musoyan V, and Cohen P 2009 *Phys. Rev.* E **79** 056102.
[18] Newman M E J 2003 *Phys. Rev.* E **68** 026121.

The bursting of housing bubble as jamming phase transition

Katsuhiro Nishinari[1], **Mitsuru Iwamura**[2], **Yukiko Umeno Saito**[3] **and Tsutomu Watanabe**[4]

[1] Research Center for Advanced Science & Technology,University of Tokyo, and PRESTO, Japan Science and Technology Corporation

[2] Waseda Business School, Waseda Univeristy

[3] Economic Research Center, Fujitsu Research Institute, and Hitotsubashi University

[4] Hitotsubashi University and Canon Institute for Global Studies

E-mail: `tknishi@mail.ecc.u-tokyo.ac.jp`

Abstract. In this paper, we have proposed a bubble burst model by focusing on transaction volume incorporating a traffic model that represents spontaneous traffic jam. We find that the phenomenon of bubble burst shares many similar properties with traffic jam formation on highway by comparing data taken from the U.S. housing market. Our result suggests that transaction volume could be a driving force of bursting phenomenon.

1. Introduction

Fluctuations in real estate prices have substantial impacts on economic activities. For example, land prices in Japan exhibited a sharp rise in the latter half of the 1980s, and its rapid reversal in the early 1990s. This large swing had led to a significant deterioration of the balance sheets of firms, especially those of financial firms, thereby causing a decade-long stagnation of the Japanese economy, which is called Japan's "lost decade". A more recent example is the U.S. housing market bubble, which started somewhere around 2000 and is now in the middle of collapsing. This has already caused substantial damages to financial systems in the U.S. and the Euro area, and it is expected that it may spread worldwide as in the case of the Great Depression in the 1920s and 30s.

These recent episodes have rekindled researchers' interest on the issue of bubbles. Economists have been regarding this phenomenon as a disorder in prices. Specifically, they define bubbles as a temporary deviation of asset prices from the corresponding fundamental values, which are basically determined by investors' expectations about future dividend stream and appropriate discount rates. Interestingly, even researchers in other areas, such as econophysics, share this empirical strategy in the sense that they look for abnormal behaviors in prices[1, 2]. However, this research strategy has overlooked an important aspect of bubbles; namely, fluctuations in asset prices tend to be closely correlated with those in the volume of transactions.

Fig.1 depicts fluctuations in housing prices and the volume of house transactions in the U.S., which shows a positive correlation between the two variables over the business cycles. More importantly, it shows that a change in transaction volume tends to lead a change in prices by one year or two[3, 4]. It is reported that similar relationships were observed for other real estate

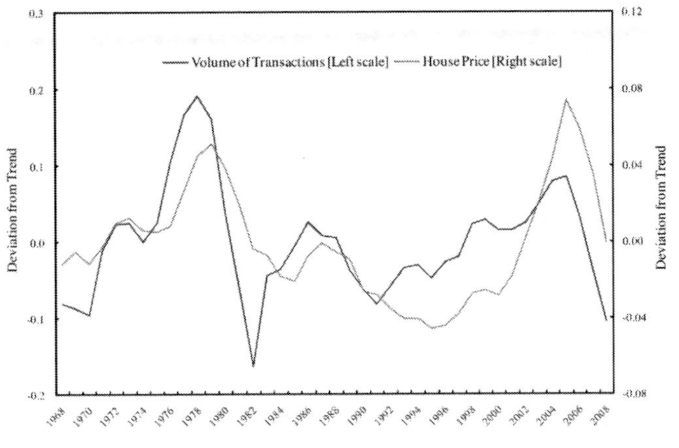

Figure 1. Fluctuations in price and transaction volume in the U.S. single-family house market. Both variables represent deviations from a linear trend.

markets in other countries, including Japan, U.K., Hong Kong, and Singapore[5, 6, 7, 8]. These evidences suggest that some sort of interaction between prices and the volume of transactions plays an important role in the process of bubble and its bursting[3, 9, 10], and that fluctuations in transaction volume, rather than those in prices, could be its driving force. Given this understanding, we focus more on transaction volume in this paper, and seek to propose a model which explains an emergence of temporary deviation of transaction volume from its appropriate level, as well as its reversal.

2. Traffic jam perspective

The idea of our modeling stems from traffic jam on highway. Each vehicle on highway usually tries to move faster if there is enough space in front of it. However, such flow becomes unstable and soon changes into congested state if the density of vehicles becomes large enough. Fig.2 is a typical observed data of flow-density relation of vehicles on highway, which is called the *fundamental diagram*. Note that flow is defined as the multiplication of density and average velocity, and density is the percentage of the occupied area of vehicles divided by the road area.

From this figure we see that flow increases according to the increase of density up to 15%, which corresponds to free flow. Instead, flow decreases above the density 20%, which correspond to jam flow. Thus we have phase transition from free to jam state at critical density 15%, and we find that free flow overlaps jam flow between the range 15%-20% in the diagram. A schematic picture of this diagram is given in Fig.3.

Free flow on $A - B$ is dangerous and unstable, since the average velocity of vehicles is high although the headway between successive vehicles is small. Thus small perturbation, e.g., weak brakes or slow down due to slope is enough to change the free flow into jamming state. Thus the branch $A - B$ is called metastable state, whose lifespan is less than 10 minutes in reality and it suddenly changes into jam flow. This scenario is indicated by an arrow in Fig.3. The existence of the metastable state is crucial for accounting so-called phantom jam, i.e., traffic jam on highway without bottlenecks. Drivers' tendency for moving faster will lead to small headway with high velocity, which eventually causes metastable flow. Then due to small perturbation traffic jam is endogenously formed spontaneously. This has been confirmed by a simple model

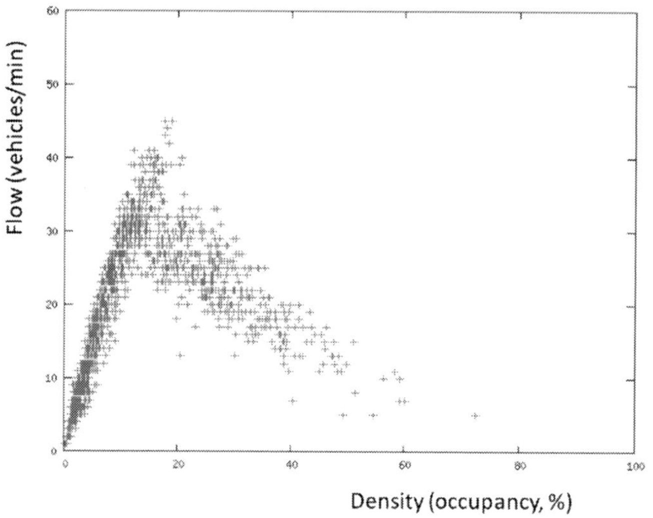

Figure 2. Typical flow-density diagram taken from Tokyo Metropolitan highway. Below (above) the critical density flow increases (decreases) according to the increase of density. Around the critical density 15% – 20% we see metastable state which corresponds unstable free flow.

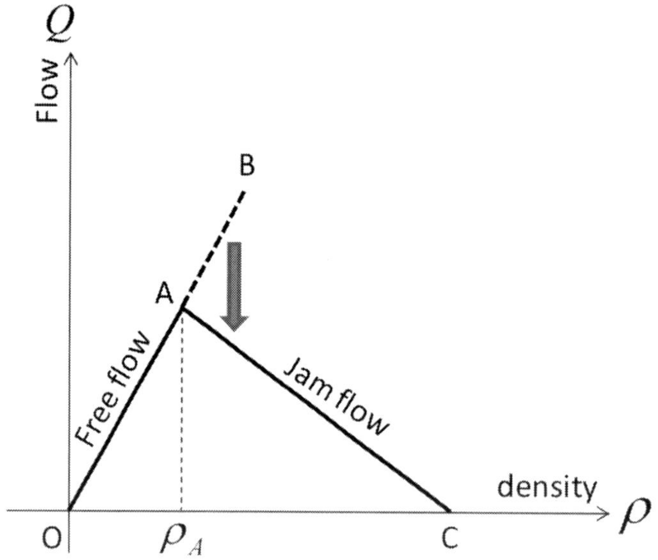

Figure 3. Schematic flow-density diagram of observed data. There is metastable state $A - B$ above the critical density ρ_A.

called the slow-to-start (SlS) model[11, 12], as well as an experiment[13].

Now we come to the point that there is similarity between spontaneous jam formation and bubble burst. Traffic jam occurs because there are too many cars relative to the available road space, while "transaction jam" in a housing market occurs because there are too many houses traded in the market relative to the total amount of liquidity (or money) supplied to the market by investors, including banks, who are outside the market. Transaction is an exchange of a house and money, and vehicle's motion is also considered as an exchange of a vehicle and a free space in front of it. Thus we propose the following analogy: For vehicle traffic,

$$\begin{aligned} \text{Density} &= \frac{\text{Space for cars}}{\text{Total available space}}, \\ \text{Average velocity} &= \frac{\text{Number of moved cars}}{\text{Number of cars}}, \end{aligned}$$

and for housing market transactions,

$$\text{Density} = \frac{\text{Housing inventory}}{\text{Total available liquidity}}, \tag{1}$$

$$\text{Average velocity} = \frac{\text{Transactions}}{\text{Housing inventory}}, \tag{2}$$

where housing inventory and transactions are evaluated as USD or other monetary unit.

Furthermore, we conjecture that metastable state plays a great role also in bubbule burst, as so in phantom jam. More precisely, we think that before the burst occurs, there is metastable transaction between real estate companies. We will show such modeling below, and compare the result with observed data taken from the US housing market.

3. Bubble burst model

3.1. Transaction and price

Before considering a model for bubble burst, let us focus on the housing sales from December 1999 to November 2008 in the U.S., during which both transaction volumes and prices showed a significant swing in our dataset. From fig.4, we immediately see that a decline in transaction volume occurred earlier than a fall in price. We also find that housing price is almost saturated around the bursting period, which is also clearly seen in the data of Japan's bubble burst in 1992. Thus in our model we focus on transaction volume instead of price in order to grasp the essence of the bursting phenomena.

Figure 4. Monthly fluctuations in average house prices and transaction volumes in the U.S. housing market from Dec 1999 to Nov 2008.

3.2. Model description

Now we propose a simple model for bubble burst incorporating the SIS model. Consider an economy with symmetric N firms dealing with houses, which are identified by i $(i = 1, \cdots, N)$ and located along a circle (Fig.5). Firm i is allowed to purchases a house from firm $i-1$, but

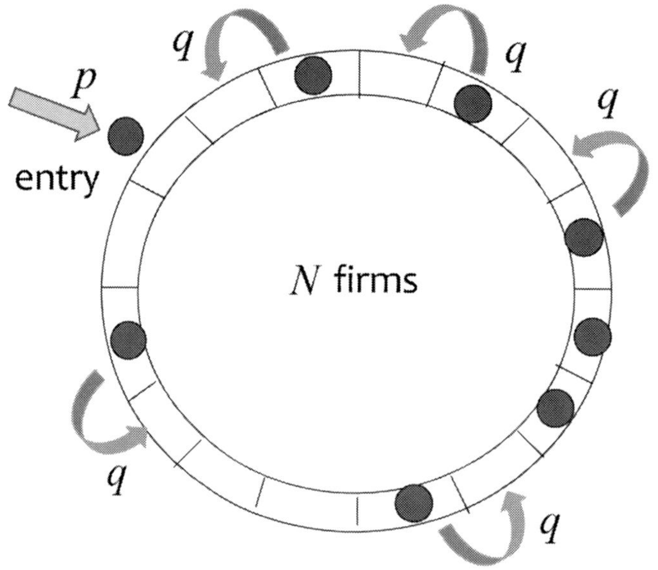

Figure 5. Bubble burst model on a circular market. There are N firms dealing houses represented by circles. The probability of successful transaction is given by q.

not allowed to do so from any other firm. For simplicity, it is assumed that the amount of cash each firm owns before buying a house is identical to the price of a house, so that each firm has its entire asset either in the form of cash or in the form of a house.

There are two rules governing transactions in this economy. First, all transactions must be in the form of an exchange of cash and a house, and no barter transactions (i.e. transactions between a house and another house) are allowed. This implies that transaction between firms i and $i-1$ never takes place in period t unless firm i holds cash at the beginning of that period. This rule corresponds to what macroeconomists call "cash-in-advance" constraint. Second, firm i becomes timid if firm i fails to make a transaction with firm $i+1$ in period t. Specifically, firm i hesitates to deal with firm $i-1$ and refuses to purchase a house in period $t+2$ even if firm i successfully has sold a house to firm $i+1$ in period $t+1$, therefore holding cash at the beginning of period $t+2$ (Fig.6). This is because, ceteris paribus, firm i is able to increase the probability that firms behind him (firm $i+1, i+2, i+3, \cdots$) hold cash, instead of a house, thereby reducing the probability that he will be refused to sell a house to firm $i+1$. This hesitation rule represents firms' preference to cash as a means to store value until the next period, because of its general acceptability.

In the modeling of vehicle traffic, there are also two important rules: the exclusion principle and the SIS rule which correspond to the above two rules in our model. Firms and houses correspond to the road cells and vehicles, respectively. The exclusion principle in vehicle traffic is that a cell can accommodate at most one vehicle for avoid collision. The SIS rule represents an inertia effect of vehicles, i.e., once a vehicle stops, then it waits extra one time step to move forward after the front cell becomes empty. In our model, this wait corresponds to the hesitation to make another transaction due to the lack of confidence about the trade partner's ability to

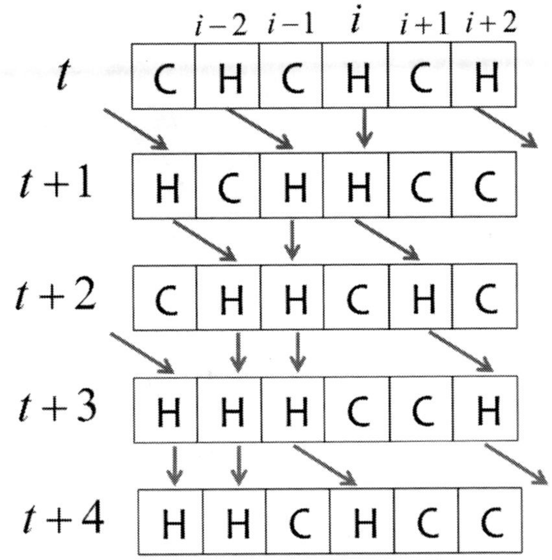

Figure 6. Hesitation of firm i in period $t+2$, which is due to the collapse of credit in the period t, leads to jam of transaction in this economy. H and C represent house and cash, respectively.

pay, and thus the fear of default. Moreover, this hesitation rule is crucial to have chain reaction of transaction fails, which leads to the bubble burst. This is similar to chain reaction of brakes on a highway under the SIS rule, which is generally observed in real traffic data.

3.3. Simulations

Given the above rules, we conduct simulations. We assume that there exists no housing inventory in this market at the beginning of period 0, but in each period, firm i, which is randomly chosen, purchases a new house from someone outside this market with probability p if (1) firm i does not hold a house (and therefore owns cash), and (2) firms ahead and behind firm i (namely, firms $i-1$ and $i+1$) do not hold a house either. In words, firm i wants to purchase a house from firm $i-1$, but cannot do that because firm $i-1$ does not own a house. At the same time, firm i expects that firm $i+1$ will be able to buy a house from him because firm i owns cash. It is only in this situation that firm i brings in a new house to the market from outside. We also assume that the probability of successful transaction between any two adjacent firms (the one with cash and the one ahead with a house) is given by q (Fig.5).

Fig.7 (left) shows fluctuations in transaction volume over time. An important thing to note is that transaction volume exhibits an abrupt decline in t=115 after a quiet period in which transaction volume is kept fairly stable. This abrupt decline is an endogenous event, and can be seen as phase transition phenomenon emerging from metastable state. Fig.7 (right) shows an example of spatiotemporal figure of transaction flow. We see jam of transaction emerges and rapidly grows endogenously, which is very similar to an abrupt change from free to congested flow in highway traffic.

The fundamental diagram is given in Fig.8. Flow is defined by multiplying eq.(1) and eq.(2). Due to the entry of houses we see linear increase of flow as density increases. The critical density is about 1/3 in this model [14], so we see the metastable transaction between 0.3 and 0.4. Then this market suddenly breaks due to the collapse of credit, and bubble bursts.

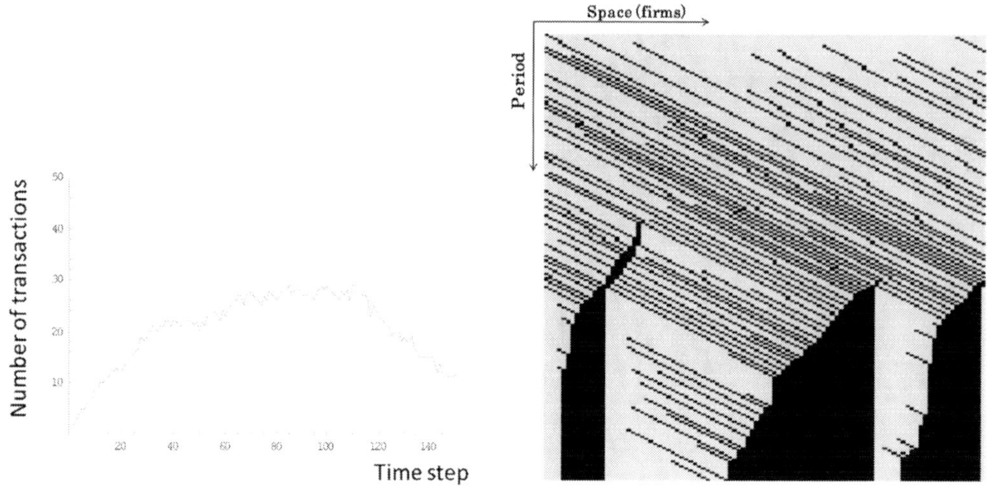

Figure 7. (left) Fluctuations in transaction volume over time. (right) Corresponding spatiotemporal figure of transaction. Small black squares represent houses which move to right due to the success of deal. Parameters N and q are set at N=100 and q=0.99 in the simulation.

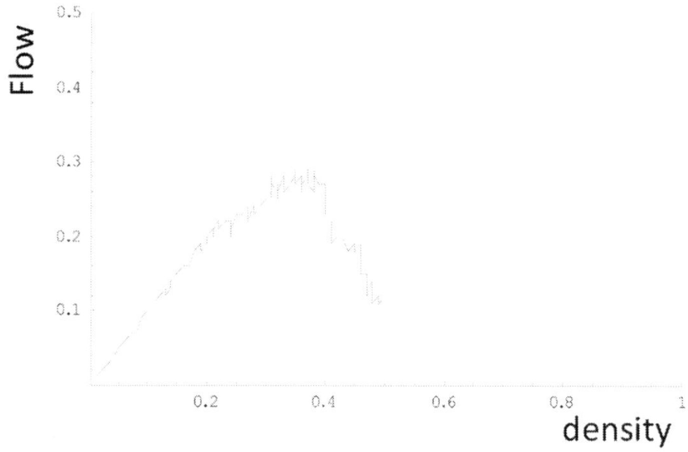

Figure 8. Flow-density diagram Parameters N and q are set at N=100 and q=0.99.

3.4. Comparison with data

The variation of transaction volume and inventory is given in Fig.9. We see that they both increase first, but at the middle of 2006 transaction begins to fall and inventory increases. Fig.10 is the fundamental diagram of this data, i.e., we depict it by taking inventory as the horizontal axis and transaction volume as the vertical axis. Because we cannot observe the total amount of liquidity supplied to the market, we make two alternative assumptions in depicting the figures. The upper one is that the total amount of liquidity changes in proportion to changes in housing prices. This represents a situation, for example, that housing prices continue to rise, and investors have an optimistic expectation about the future course of prices, and thus willingly provide additional liquidity to the market. This assumption is adopted in the upper figure where

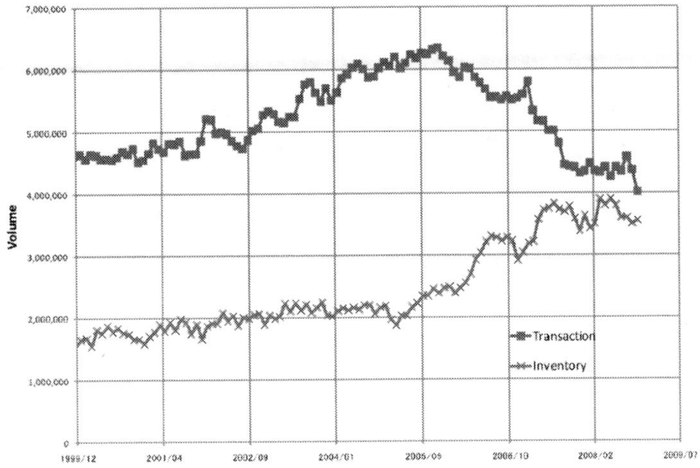

Figure 9. Transaction volume and inventory in the U.S. housing market from Dec 1999 to Nov 2008.

transaction and inventory are both measured in the number of housing units. Alternatively, we may assume that the total liquidity supplied to the market from the outside investors is exogenously determined (i.e. it does not depend on the evolution of prices); more specifically, we assume that the total amount of liquidity does not change at all during the entire sample period. This assumption is adopted in the lower figure where transaction and inventory are evaluated in dollars. The two figures should be compared with Fig.8, showing similar properties such as increases of flow at early stage and its collapse due to the decrease of transaction.

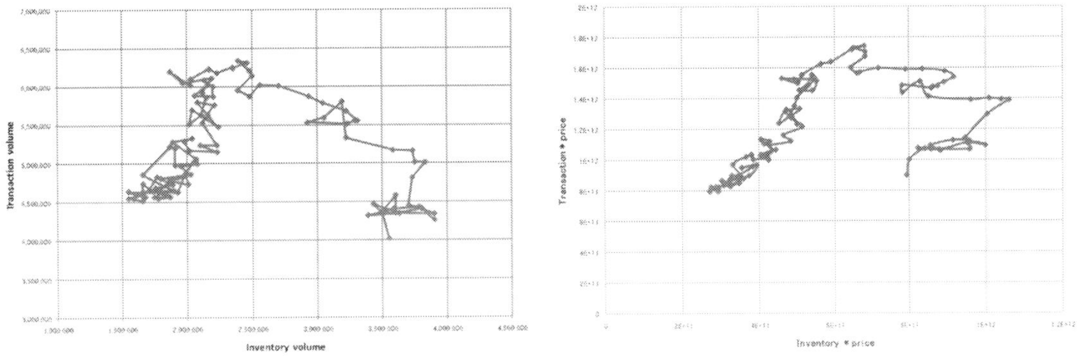

Figure 10. Fundamental diagram, i.e., inventory versus transaction volume depicted from the data shown in Fig.9. The horizontal and vertical axes represent the number of housing units in the left figure and dollars in the right figure.

4. Concluding discussions

In this paper, we have proposed a bubble burst model by using a traffic model that represents spontaneous traffic jam. We find that the phenomenon of bubble burst shares many similar properties with traffic jam formation. Especially we would like to stress on the importance of the SIS rule which will be recast as hesitation in our model. The simulation results obtained in our model is similar to those of the data taken from the U.S. housing market. We focus on transaction in our model instead of the price, and this has not been considered up to now in studying bubble burst phenomena. Our result suggests that the transaction could be a driving force of bursting phenomenon.

In our theoretical model we know the value of the critical density, then we can predict whether the market is now in a dangerous metastable state or not by checking the amount of inventory. Of course we don't know this critical value in the real market, we hope that our analysis will help to judge the danger and avoid the burst and crash of market. Studying other bubble phenomena, e.g., oil market and extensions of our model such as introducing price change and various dealing network are ongoing and future works.

References

[1] Zhou W X and Sornette D 2006 *Physica A* **361** 297
[2] Watanabe K, Takayasu H and Takayasu M 2007 *Physica A* **383** 120
[3] Stein J 1995 *Quarterly Journal of Economics* **110** 379
[4] Wheaton W C and Lee N J 2009 *Working Paper, Available at* http://econ-www.mit.edu/files/3781
[5] Ortalo-Magne F and Rady S 2004 *Journal of Housing Economics* **13** 287
[6] Andrew M and Meen G 2003 *Real Estate Economics* **31** 99
[7] Leung C K Y, Lau G C K and Leong Y C F 2002 *Journal of Real Estate Research* **23** 253
[8] Edelstein R H and Lum S K 2004 *Journal of Housing Economics* **13** 342
[9] Genesove D and Mayer C 2001 *Quarterly Journal of Economics* **116** 1233
[10] Caplin A and Leahy J 2008 *NBER Working Paper* 14605
[11] Takayasu M and Takayasu H 1993 *Fractals* **1** 860
[12] Nishinari K and Takahashi D 2000 *J. Phys. A: Math. Gen.* **33** 7709
[13] Sugiyama Y, Fukui M, Kikuchi M, Hasebe K, Nakayama A, Nishinari K, Tadaki S and S Yukawa 2008 Traffic jams without bottlenecks *New. J. Phys.* **10** 033001
[14] Sakai S, Nishinari K and IIda S 2006 *J.Phys.A: Math. Gen.* **39** 15327

The difference of growth rate distributions between sales and profits

Atushi Ishikawa[1], Shouji Fujimoto[1] and Masashi Tomoyose[2]
[1]Kanazawa Gakuin University, 920-1392, Japan
[2]University of the Ryukyus, 903-0213, Japan

E-mail: ishikawa@kanazawa-gu.ac.jp

Abstract. Using numerical simulations, the authors exhibit the difference between two types of the growth rate distributions, the one of which is observed in both positive and negative data such as profits, and the other of which is in non–negative data such as sales. In the simulation, firstly the Langevin equation generates both positive and negative variables, the growth rate distributions of which are linear functions of the logarithmic growth rate. By superposing the variables not to be negative, we find that the growth rate distributions of the non–negative variables have wider tails than line shape on a log–log scale. At the same time, two types of Non–Gibrat's Laws in the middle scale range are also confirmed as observed in real economic data.

1. Introduction

In order to understand economic and social systems, it is important to study collective behavior of the elements. One of remarkable examples is Pareto's Law [1]:

$$P(v) \propto v^{-(\mu+1)} \quad \text{for} \quad v > v_{\rm th} \ , \tag{1}$$

which is observed in economic data such as assets, sales, profits, income of firms, the number of employees and personal income (denoted by v) over some size threshold $v_{\rm th}$. Here, $P(v)$ is the probability density function (pdf) and the power μ is called Pareto index. It is important to understand the mechanism generating Pareto's Law because a small number of firms in the power–law range dominate a large percent of the overall sales, profits and so forth. Pareto' Law has been well investigated by using various models ([2]–[5] for instance). Recently, by using no model, it has been shown that Pareto's Law can be explained by laws observed in a massive amount of digitized economic data. Fujiwara et al. [6, 7] point out that Pareto's Law can be derived from the Law of Detailed Balance and from Gibrat's Law [8].

Detailed Balance is time-reversal symmetry:

$$P_{12}(v_1, v_2) = P_{12}(v_2, v_1) \tag{2}$$

as observed in the equilibrium system which corresponds to a stable economy. Here, firm sizes at the two successive points in time are denoted by v_1 and v_2, and $P_{12}(v_1, v_2)$ is the joint pdf. The growth rate R is defined as the ratio $R = v_2/v_1$. Gibrat's Law states that the conditional pdf of the growth rate $Q(R|v_1)$ is independent of the initial value v_1 as follows:

$$Q(R|v_1) = Q(R) \ , \tag{3}$$

where $Q(R|v_1)$ is defined as $Q(R|v_1) = \frac{P_{1R}(v_1,R)}{P(v_1)}$ by using the pdf $P(v_1)$ and the joint pdf $P_{1R}(v_1, R)$.

In these analyses of economic data, two types of the growth rate distributions are observed. For instance in Ref. [9], it is reported that the shape of the growth rate distribution of profits is different from that of sales by employing Japanese firms data [10]. The growth rate distribution of profits can be approximated by linear functions of the logarithmic growth rate $r = \log_{10} v_2/v_1$ as follows:

$$\log_{10} q(r) = c - t_+ \, r \quad \text{for } r > 0 \,, \tag{4}$$
$$\log_{10} q(r) = c + t_- \, r \quad \text{for } r < 0 \,. \tag{5}$$

Here, $q(r)$ is the pdf of r, and is related to that of R by $\log_{10} q(r) = \log_{10} Q(R) + r + \log_{10}(\ln 10)$. The linear approximation is, however, not suitable for the growth rate distribution of sales because the distribution has a curvature and wider tails than line shape on a log–log scale.

This difference has been observed in other literature employing not only Japanese firms data but also European and North American firms data (see [11]–[14] for instance). This aspect has been also observed in other economic data. Data having the growth rate distribution with no wide tail are income of firms [12, 15, 9], and GDP [16] for instance. In contrast, data having the growth rate distribution with wide tails are personal income in Japan [6], and assets and sales in France and the number of employees in the UK [7] for instance.

From these observations, we have made a simple hypothesis that the difference between shapes of growth rate distributions comes from being subtracted or not [9]. On the one hand, for example, profits are calculated by a subtraction of total expenditure from total sales at a rough estimate. The values can be both positive and negative. On the other hand, sales are not calculated by any subtraction. The values cannot be negative. This hypothesis can explain above observations in the literatures. On one hand, profits, income of firms and GDP are calculated by a subtraction, therefore, the growth rate distributions have no wide tail. The growth rate distributions are approximated by linear functions of r on a log–log scale. On the other hand, assets, sales, the number of employees and personal income are not calculated by any subtraction, therefore, the growth rate distributions have wide tails. The linear approximation is not applicable for the growth rate distributions.

In order to verify this hypothesis, one of the authors (A. I) has studied temporal changes of high–sales in Japan and observed that the growth rate distributions of temporal sales changes have no wide tail as expected [9]. This confirms our hypothesis that the growth rate distributions of both positive and negative variables are approximated by linear functions of r and that those of non–negative variables have wider tails than line shape on a log–log scale.

This hypothesis is not restricted in the large scale range. It is well known that Pareto's Law (1) breaks below the size threshold $v_{\rm th}$ [17, 18]. It is also important to study the distribution in the middle scale range because the majority of firms are middle-sized. In the middle size range, one of the authors (A. I) has showed that the log–normal distribution:

$$P(v) = C \, v^{-(\mu+1)} \, e^{-\alpha \ln^2 \frac{v}{v_{\rm th}}} \quad \text{for} \quad v < v_{\rm th} \tag{6}$$

can also be deduced using Detailed Balance and First Non–Gibrat's Law [19] along works in Ref. [6, 7]. Non–Gibrat's Law, observed below the size threshold $v_{\rm th}$, describes the dependence of the growth rate distribution on the initial value. In Ref. [20], the following is proved. If the growth rate distribution is described by linear functions of r even in the middle scale range as

$$\log_{10} q(r|v_1) = c(v_1) - t_+(v_1) \, r \quad \text{for } r > 0 \,, \tag{7}$$
$$\log_{10} q(r|v_1) = c(v_1) + t_-(v_1) \, r \quad \text{for } r < 0 \,, \tag{8}$$

under Detailed Balance, the dependence of the initial value is uniquely fixed as follows:

$$t_{\pm}(v_1) = t_{\pm}(v_{\rm th}) \pm \alpha \ \ln \frac{v_1}{v_{\rm th}} \ . \tag{9}$$

This is verified in empirical data [20]. The probability of positive growth decreases and the probability of negative growth increases as the initial value v_1 increases in the middle scale range. In this paper, we designate Eqs. (7)–(9) First Non-Gibrat's Law.

For the growth rate distributions of assets or sales in the middle scale range, other type of Non-Gibrat Law is reported [21]–[23]. The probability of positive and negative growth decreases as the initial value increases. Let this law be designated by Second Non-Gibrat's Law. Let us make another hypothesis that the difference between two shapes of the growth rate distributions is related to the difference between two Non-Gibrat's Laws in the middle scale range.

In this study, we clarify these two hypotheses by using numerical simulations as follows:

(i) We generate both positive and negative variables, the growth rate distributions of which are linear functions of r. The variables follow Pareto's Law in the large scale range, and follow the log–normal distribution in the middle scale range. In the middle scale range, the growth rate distributions follow First Non–Gibrat's Law.

(ii) By superposing the variables not to be negative, we find that the growth rate distributions of the non–negative variables have wider tails than line shape on a log–log scale.

(iii) We also find that the growth rate distributions of the non–negative variables follow Second Non–Gibrat's Law in the middle scale range.

2. Empirical Observations

In this section, we present the results of empirical analyses of our previous studies in order to set up the simulation model. We have employed the database "CD Eyes 50" published by TOKYO SHOKO RESEARCH, LTD [10]. In the database, top 500 thousand sales data of Japanese firms are available. This database is thought to be approximately exhaustive in the large scale range. The database also includes both positive and negative profits data. The number of firms with positive profits is approximately 300 thousand and that with negative profits is approximately 40 thousand. The remaining approximately 160 thousand data do not include profits data. We have excluded firms with negative profits, the number of which is significantly less than that with positive profits. This is due to the fact that the negative profits data gathered from high–sales data are exclusive. Despite the fact that we do not have a complete picture of the positive profits data for firms in the middle scale range, we have decided to employ this data to investigate the consistency of the laws with the data.

Hereafter, let "non–negative variables" and "both positive and negative variables" be denoted by x and v, respectively. Figure 1 shows the scatter plots of profits and sales in the database. In each scatter plot, Detailed Balance (2) has been approximately confirmed by applying the one-dimensional (1–dim) Kolmogorov–Smirnov (K–S) test [9].

Firstly, let us analyze profits data. In order to identify Gibrat's Law in the large scale range and Non–Gibrat's Law in the middle scale range, we divide the range of the initial profits v_1 into logarithmically equal bins as $v_1 \in [10^{2+0.2(n-1)}, 10^{2+0.2n})$ $(n = 1, 2, \cdots, 20)$. The conditional pdfs of the profits growth rate are shown in Fig. 2. In this figure, we observe that the growth rate distributions can be approximated by linear functions of r in the range for $n \geq 6$ $(v_1 \geq 10^3)$. By applying the linear approximations (7) and (8) to the data in Fig. 2, the relation between v_1 and $t_{\pm}(v_1)$ is obtained (Fig. 3). On the one hand, Fig. 3 shows that $t_{\pm}(v_1)$ hardly responds to v_1 for $n = 16, \cdots, 20$ $(v_1 \geq 10^5)$. This means that Gibrat's Law (3) or (4)–(5) is valid in the large scale range. On the other hand, $t_+(v_1)$ linearly increases and $t_-(v_1)$ linearly decreases symmetrically with $\log_{10} v_1$ for $n = 6, \cdots, 10$ $(10^3 \leq v_1 < 10^5)$. This is First Non-Gibrat's Law (7)–(9), which

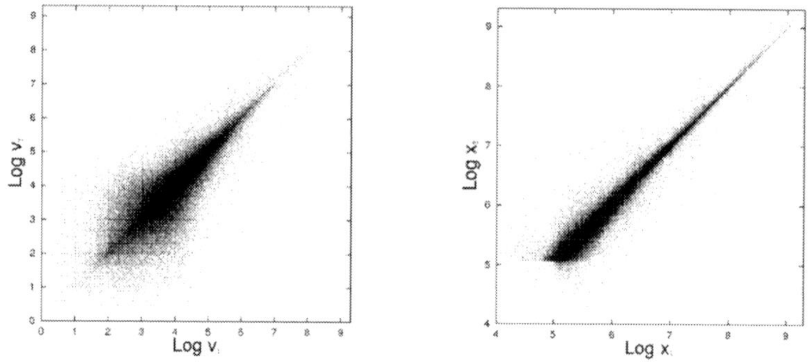

Figure 1. The scatter plots of profits and sales in the database. Here, v_1 and v_2 (x_1 and x_2) represent profits (sales) in 2004 and 2005, respectively. The number of profits and the number of sales are 177, 492 and 406, 385 sets, respectively. In each case, Detailed Balance (2) is approximately confirmed by applying the 1-dim K–S test.

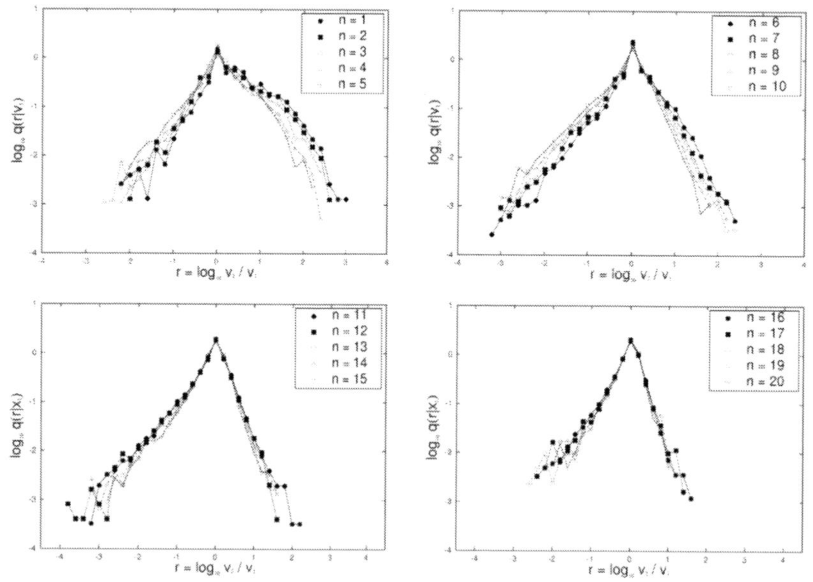

Figure 2. Conditional pdfs $q(r|v_1)$ of the log profits growth rate $r = \log_{10} v_2/v_1$ from $2004(v_1)$ to $2005(v_2)$. The data points are classified into twenty bins of the initial profits with equal magnitude in logarithmic scale, $v_1 \in [10^{2+0.2(n-1)}, 10^{2+0.2n})$ $(n = 1, 2, \cdots, 20)$ thousand yen.

can be derived analytically from the linear approximations (7) and (8) under Detailed Balance (2) [20]. In Fig. 3, the parameters are estimated as follows: $\alpha \sim 0$ for $v > v_{\rm th}$ (the large scale range), $\alpha \sim 0.14$ for $v_{\rm min} < v < v_{\rm th}$ (the middle scale range), $t_+(v_{\rm th}) \sim 2$, $t_-(v_{\rm th}) \sim 1$, $v_{\rm th} \sim 10^{2+0.2(16-1)} = 10^5$ thousand (= 100 million) yen and $v_{\rm min} \sim 10^{2+0.2(6-1)} = 10^3$ thousand (= 1 million) yen. Rigorously, a constant parameter α must not take different values. In the database, however, a large number of firms stay in the same range for two successive years. This parameterization is approximately valid for describing the pdf.

Figure 4 shows that the resultant pdf (6) behaves as Pareto's Law (1) in the large scale range where Gibrat's Law is hold ($\alpha \sim 0$), and behaves as the log–normal distribution in the middle scale range where First Non–Gibrat's Law is hold ($\alpha \sim 0.14$).

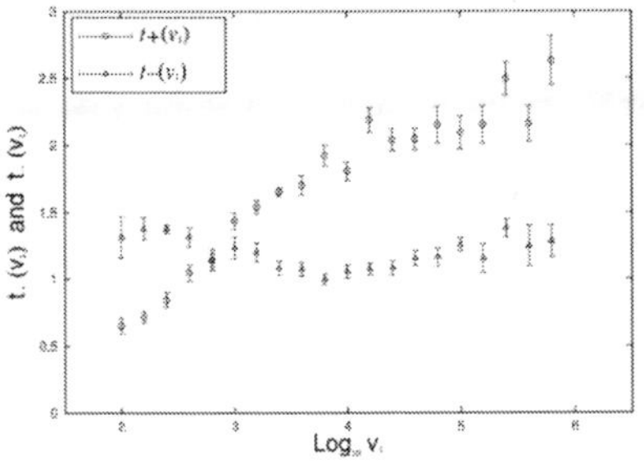

Figure 3. The relation between the lower bound of each bin $v_1 \in [10^{2+0.2(n-1)}, 10^{2+0.2n})$ and $t_\pm(v_1)$. From the left, each point on the graph represents $n = 1, 2, \cdots, 20$. The values are measured by the least square method in the region $0 \le |r| \le 2$ in Fig. 2.

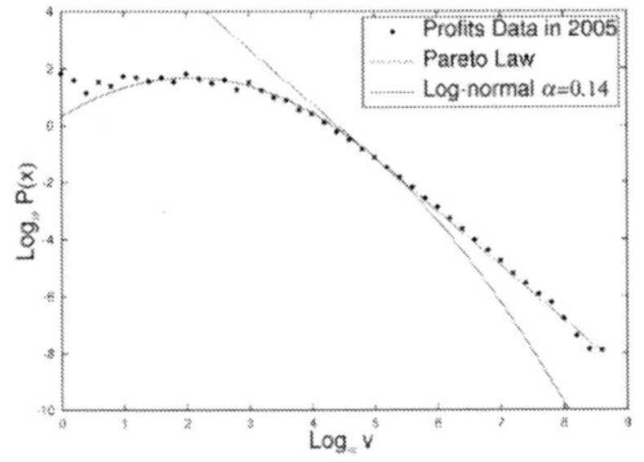

Figure 4. The resultant pdf of positive profits (6) behaves as Pareto's Law (1) in the large scale range where Gibrat's Law is hold ($\alpha \sim 0$), and behaves as the log–normal distribution in the middle scale range where First Non–Gibrat's Law is hold ($\alpha \sim 0.14$)

Secondly, we analyze sales data. In the same manner as in profits data analysis, we divide the range of the initial sales x_1 into logarithmically equal bins as $x_1 \in [10^{6+0.2(n-1)}, 10^{6+0.2n})$ $(n = 1, 2, \cdots, 10)$. The conditional pdfs of the sales growth rate are shown in Fig. 5. Note that the database does not include sales data in the middle scale range. Therefore, only Gibrat's Law (3) is observed. In Fig. 5, the growth rate distributions hardly change as the initial sales value x_1 increases. In this case, Pareto's Law (1) is induced from Gibrat's Law (3) and from Detailed Balance (2) [6, 7]. Figure 6 shows that the sales data follow Pareto's Law (1) in the large scale range. We should not pay attention the break of Pareto's Law because this comes from the exhaustiveness of sales in the database.

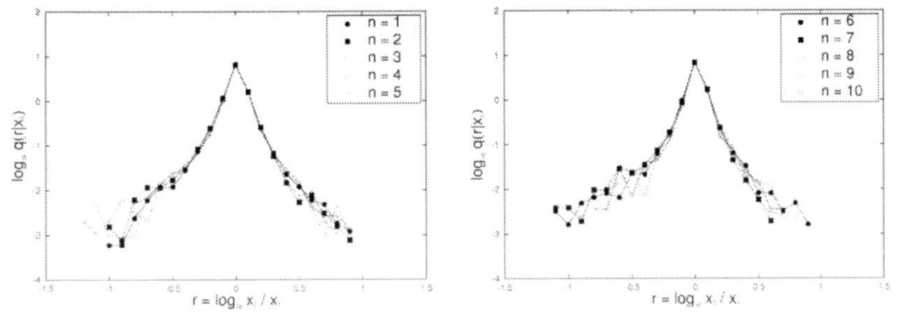

Figure 5. Conditional pdfs $q(r|x_1)$ of the log sales growth rate $r = \log_{10} x_2/x_1$ from $2004(x_1)$ to $2005(x_2)$. The data points are classified into ten bins of the initial sales with equal magnitude in logarithmic scale, $x_1 \in [10^{6+0.2(n-1)}, 10^{6+0.2n})$ $(n = 1, 2, \cdots, 10)$ thousand yen.

As stated in introduction, the difference of growth rate distributions between sales and profits exists. Most distinguishing feature is that the growth rate distributions of profits are approximated by linear functions of r ($n \ge 6$ in Fig. 2), however, those of sales are not (Fig. 5). This feature is also observed in existing literatures employing various economic data in many

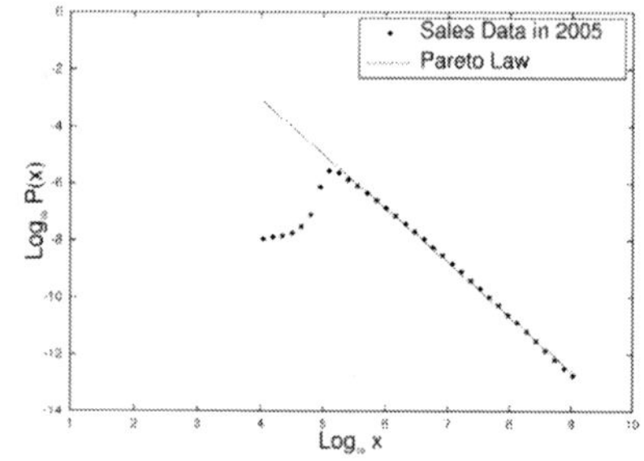

Figure 6. The sales data in the database follow Pareto's Law in the large scale range. This database does not include middle-seized sales data. The break of Pareto's Law comes from the exhaustiveness of sales in the database. We should not pay attention the break.

countries ([11]–[16] for instance). We have made a simple hypothesis that the difference between shapes of growth rate distributions comes from being subtracted or not [9].

In order to confirm this hypothesis, one of the authors (A. I) has studied temporal changes of high–sales in Japan [9]. Here, let high–sales at the three successive points in time be denoted by x_0, x_1 and x_2. We consider two temporal changes $x_1 - x_0 \equiv v_1$ and $x_2 - x_1 \equiv v_2$, which can be both positive and negative. The data are classified into the following four quadrants: $(v_1 > 0, v_2 > 0)$, $(v_1 < 0, v_2 > 0)$, $(v_1 < 0, v_2 < 0)$ and $(v_1 > 0, v_2 < 0)$. In each quadrant, distributions in the growth rate of temporal sales changes $R = |v_2/v_1|$ are shown in Fig. 7. In four cases, no wide tail is observed as expected. It is possible to compare growth rate distributions of profits with those of temporal sales changes only in first quadrant because exhaustive negative profits data are not available. These two growth rate distributions are same kind in the sense that they have no wide tail. This confirms our hypothesis that the growth rate distributions of both positive and negative variables are approximated by linear functions of r and that those of non–negative variables have wide tails on a log–log scale. At the same time, Detailed Balance and First Non–Gibrat's Law have been also approximately observed in four quadrants.

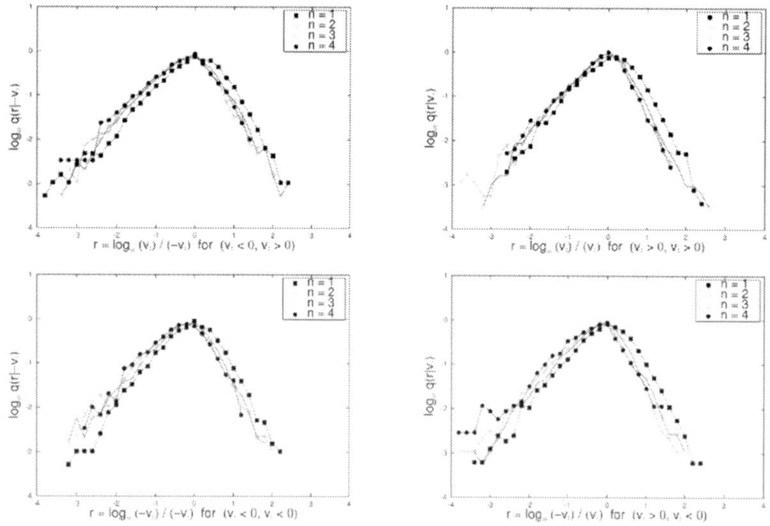

Figure 7. Conditional pdfs $q(r||v_1|)$ in the log growth rate of the temporal sales change $r = \log_{10}|v_2/v_1|$ for cases $(v_1 < 0, v_2 > 0)$, $(v_1 > 0, v_2 > 0)$, $(v_1 < 0, v_2 < 0)$ and $(v_1 > 0, v_2 < 0)$. In each figure, data points are classified into four bins of the initial temporal sales change with equal magnitude in logarithmic scale, $|v_1| \in [10^{4.5+0.5(n-1)}, 10^{4.5+0.5n})$ $(n = 1, 2, \cdots, 4)$ thousand yen. Here, $v_2 = x_2 - x_1$ is the change between 2004 (x_1) and 2005 (x_2), and $v_1 = x_1 - x_0$ is between 2003 (x_0) and 2004 (x_1).

Unfortunately, the database we have employed does not include sales data in the middle scale range. By following the reports [21]–[23], Second Non–Gibrat's Law, that the probability of positive and negative growth decreases as the initial value increases, is observed in the middle scale range. First and Second Non–Gibrat's Laws are depicted in Figs. 8 and 9, respectively.

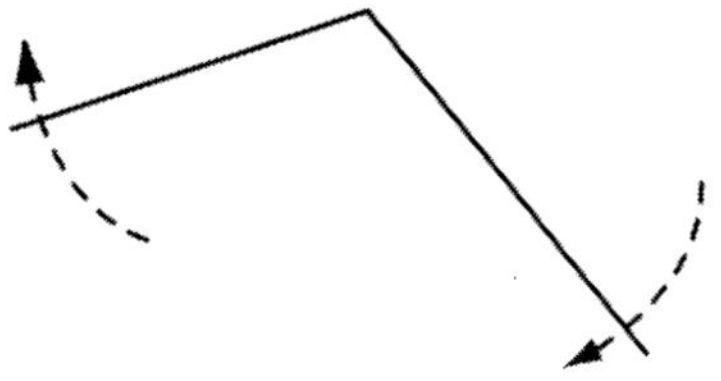

Figure 8. First Non–Gibrat's Law: The horizontal axis is the logarithm of the growth rate (r) and the vertical axis is the logarithm of its pdf ($\log_{10} q(r|v_1)$). The dashed lines with arrow means the change of the growth rate distribution as the initial profits v_1 increases in the middle scale range. This corresponds to the case $6 \leq n \leq 10$ in Fig. 2.

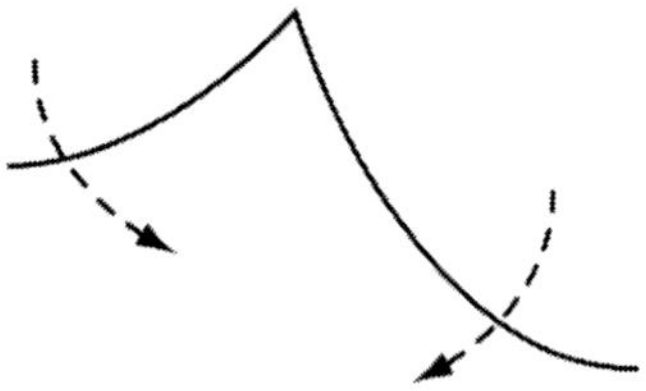

Figure 9. Second Non–Gibrat's Law: The horizontal axis is the logarithm of the growth rate (r) and the vertical axis is the logarithm of its pdf ($\log_{10} q(r|x_1)$). The dashed lines with arrow means the change of the growth rate distribution as the initial sales x_1 increases in the middle scale range.

3. TFI Simulation Model

In Ref. [24], based on the work by Takayasu et al. [5], the authors have proposed the simulation model which leads to Pareto's Law in the large scale range where Gibrat's Law is hold, and to the log-normal distribution in the middle scale range where First Non–Gibrat's Law is hold. The model is given by the Langevin equation

$$v(t+1) = b(v(t), t)v(t) + f(t) , \tag{10}$$

where $b(v(t), t)$ is a non–negative multiplicative stochastic noise and $f(t)$ is a random additive noise.

In the simulation, we adopt the distribution of $b(v(t), t)$ to be Eqs. (7)–(9) with $\alpha = 0$ for $v > v_{\rm th}$ (Gibrat's Law), $\alpha \neq 0$ for $v_{\rm int} < v < v_{\rm th}$ (First Non–Gibrat's Law) and $\alpha = 0$ for $v < v_{\rm int}$ keeping $t_{\pm}(x)$ continuously (Fig. 10).[1] Figure 10 corresponds to Fig. 3 observed in empirical profits data. As the middle scale range in which First Non–Gibrat's Law is hold, we observe the range $\max(v_{\rm min}, v_{\rm int}) < v < v_{\rm th}$. Here, $v_{\rm int}$ is depicted in Fig. 10 and $v_{\rm min}$ is defined as the bottom bound of the range where Detailed Balance is observed. In this paper, we refer to the combination of Gibrat's Law in the large scale range and Non–Gibrat's Law in the middle scale range as Extended-Gibrat's Law. We also adopt the distribution of $f(t)$ to be Weibull distribution: $\frac{k}{\lambda}(\frac{x}{\lambda})^{k-1}\exp[-(\frac{x}{\lambda})^k]$.

The typical scatter plot of the simulation is shown in Fig. 11, where we take $t_+(v_{\rm th}) = 2.5$, $t_-(v_{\rm th}) = 1.5$, $v_{\rm th} = 10^6$, $k = 0.5$, $\lambda = 30$ and $\alpha = 0.1$. In order to claim that this model is consistent with the empirical observation, Detailed Balance and Extended–Gibrat's law must be satisfied. From Eq. (10), Extended–Gibrat's law (7)–(9) is hold in the range where a noise $f(t)$ is negligible. Detailed Balance (2) is also confirmed. The validity is not rejected within the 5% significance level by applying the 1-dim K–S test in the range approximately over $v_{\rm min} = 10^4$ (Fig. 12). It is worth noting that Detailed Balance is not imposed in this simulation a priori.

[1] The last parameterization is imposed to exclude immoderate slopes of growth rate distributions.

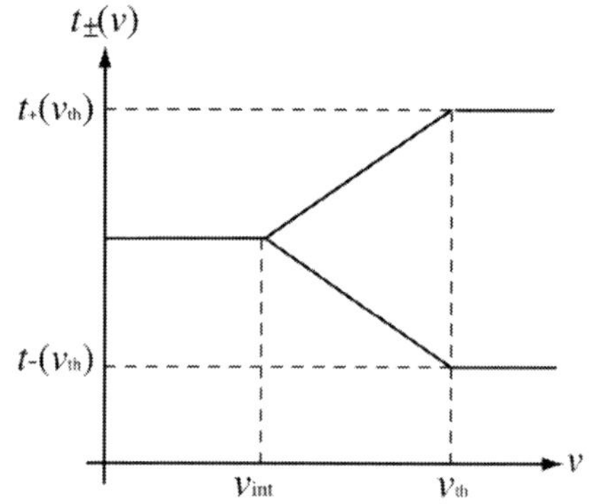

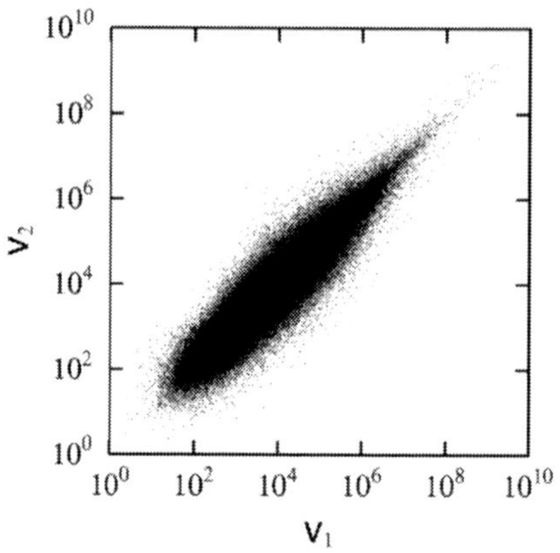

Figure 10. Continuous functions $t_{\pm}(v)$. The horizontal axis is logarithmic scale.

Figure 11. The scatter plot of data points under Extended–Gibrat's Law. The number of data points is 500,000.

The resultant pdf of v is shown in Fig. 13, where the log-normal distribution in the middle scale range is observed in addition to Pareto's Law in the large scale one. From these results, we recognize that this simulation model fits with the empirical observations.

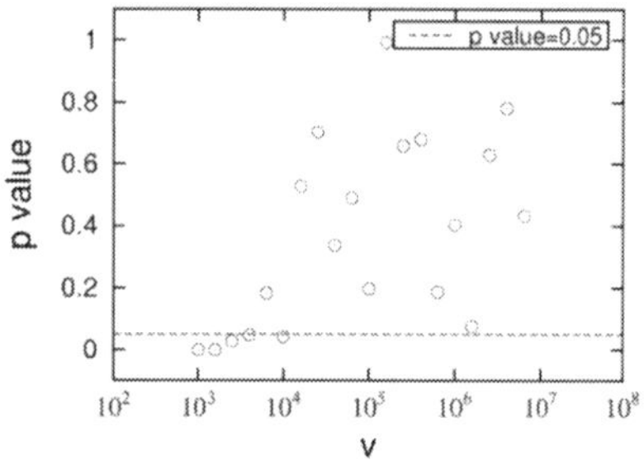

Figure 12. Each p value of the 1–dim K–S test for the scatter plot in Fig. 11. We compared the distribution sample for $P(v_1 \in [10^{3+0.2(n-1)}, 10^{3+0.2n}), v_2)$ with another sample for $P(v_1, v_2 \in [10^{3+0.2(n-1)}, 10^{3+0.2n}))$ $(n = 1, 2, \cdots, 20)$. Detailed Balance is verified in the range approximately over 10^4 .

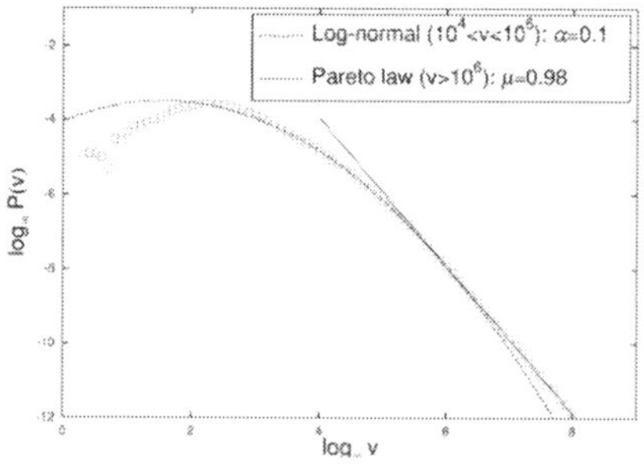

Figure 13. The resultant pdf $P(v)$. In the large scale range ($v > 10^6$), Pareto's Law is observed. In the middle scale one ($10^4 < v < 10^6$), the log–normal distribution is observed.

4. Simulation Model for Profits and Sales

The simulation in the previous section generates the stochastic variables v, the growth rate distributions of which are adopted to be linear functions of r (7) and (8). The multiplicative noise b and the additive one f are both positive, therefore, the stochastic variables v are also positive. However, in the case that the growth rate distributions have no wide tail, the stochastic variables v should be both positive and negative in our hypothesis.

From the results of analyses in temporal changes of sales data [9], we extend the model to generate Detailed Balance and First Non-Gibrat's Law in the second, third and fourth quadrants, as observed in the first quadrant. This can be realized by the modified Langevin equation [25]:

$$v(t+1) = a(t)b(v(t),t)v(t) + f(t) \ , \tag{11}$$

where $a(t) = 1$ with probability 0.5 and $a(t) = -1$ with probability 0.5. In this case, the distribution of f is adopted to be Gaussian $N(0, \sigma^2)$. The typical scatter plot of the simulation is shown in Fig. 14, where we take $t_+(v_{\rm th}) = 2.6$, $t_-(v_{\rm th}) = 1.4$, $v_{\rm th} = 10^6$, $\sigma = 20$ and $\alpha = 0.2$. In each quadrant, Detailed Balance is approximately confirmed by applying the 1–dim K–S test (Fig. 15). Even in modified Langevin Eq. (11), Extended–Gibrat's Law (7)–(9) is hold in the range where a noise $f(t)$ is negligible.

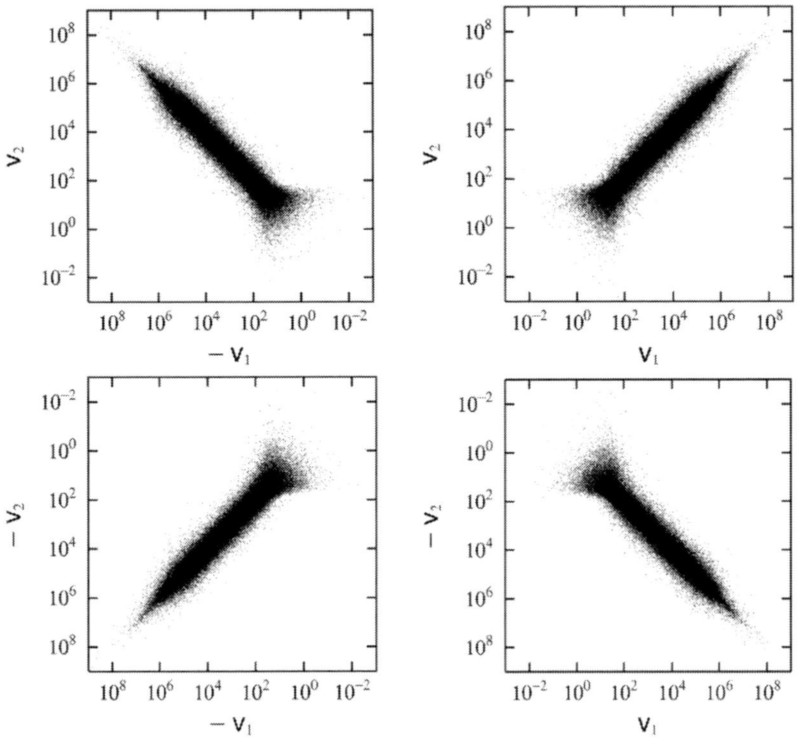

Figure 14. The scatter plots of data points in four quadrants. The total number of data points is 500,000.

Using these both positive and negative variables v, we obtain non–negative variables x by assuming the change of x to be v:

$$v(t) \ = \ x(t+1) - x(t) \ . \tag{12}$$

In the next simulation, non–negative variables x are updated iteratively with the restriction $x > 0$ as follows:

$$x(t+1) = \begin{cases} x(t) + v(t) & \text{if} \quad x(t+1) > 0, \\ x(t) & \text{if} \quad x(t+1) < 0. \end{cases} \tag{13}$$

The resulting scatter plot of x is shown in Fig. 16. Detailed Balance is approximately verified by applying the 1–dim K–S test (Fig. 17).

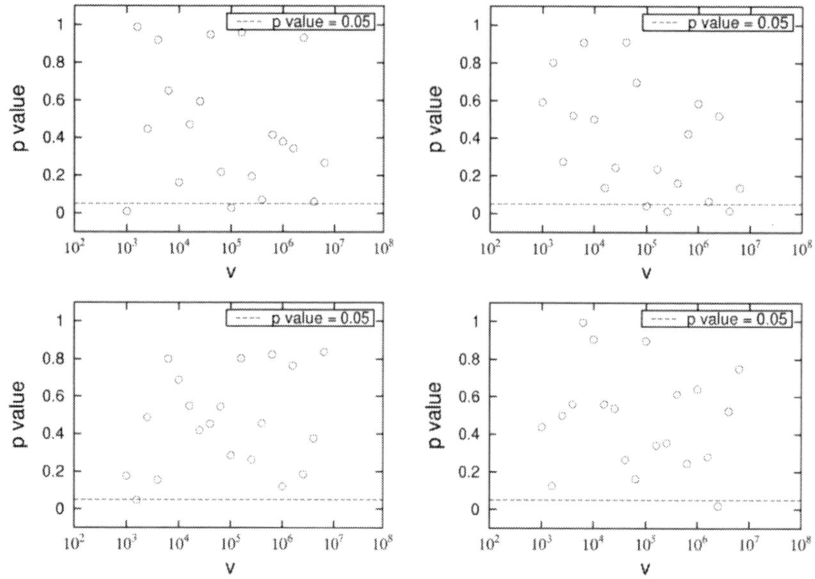

Figure 15. P values of the 1–dim K–S test for each scatter plot in Fig. 14. We compared the distribution sample for $P(|v_1| \in [10^{3+0.2(n-1)}, 10^{3+0.2n}), |v_2|)$ with another sample for $P(|v_1|, |v_2| \in [10^{3+0.2(n-1)}, 10^{3+0.2n}))$ $(n = 1, 2, \cdots, 20)$. In each quadrant, Detailed Balance is approximately verified.

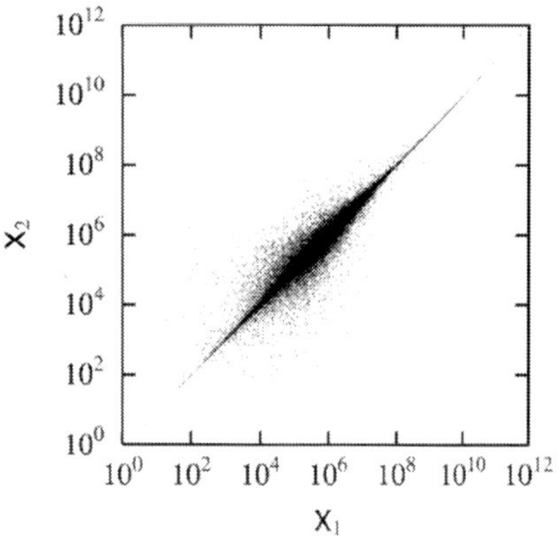

Figure 16. The scatter plot of non–negative data points x, the number of which is 500,000.

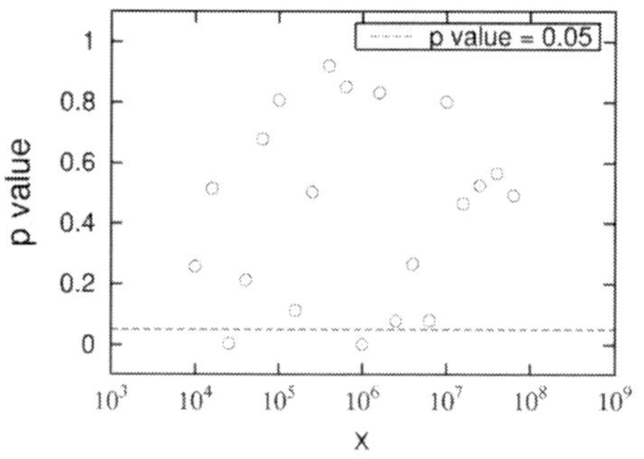

Figure 17. P values of the 1–dim K–S test for the scatter plot in Fig. 16. We compared the distribution sample for $P(x_1 \in [10^{4+0.2(n-1)}, 10^{4+0.2n}), x_2)$ with another sample for $P(x_1, x_2 \in [10^{4+0.2(n-1)}, 10^{4+0.2n}))$ $(n = 1, \cdots, 20)$. Detailed Balance is approximately confirmed.

The main result in this paper is the growth rate distributions of non–negative variables x shown in Fig. 18. The point is that the growth rate distributions have wide tails. In addition, Second Non–Gibrat's Law is also observed in the middle scale range. The probability of positive and negative growth decreases as the bin size of the classification of x increases as expected.

The resultant cumulative pdf of x is shown in Fig. 19.

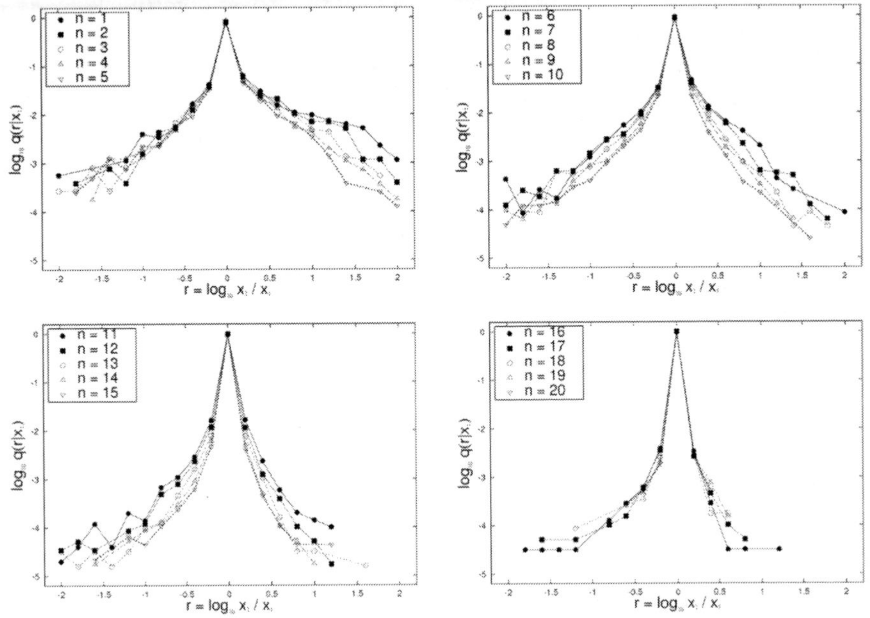

Figure 18. Conditional pdfs $q(r|x_1)$ of the log growth rate $r = \log_{10} x_2/x_1$. The data points are classified into twenty bins of the initial variables with equal magnitude in logarithmic scale, $x_1 \in [10^{4+0.2(n-1)}, 10^{4+0.2n})$ $(n = 1, 2, \cdots, 20)$.

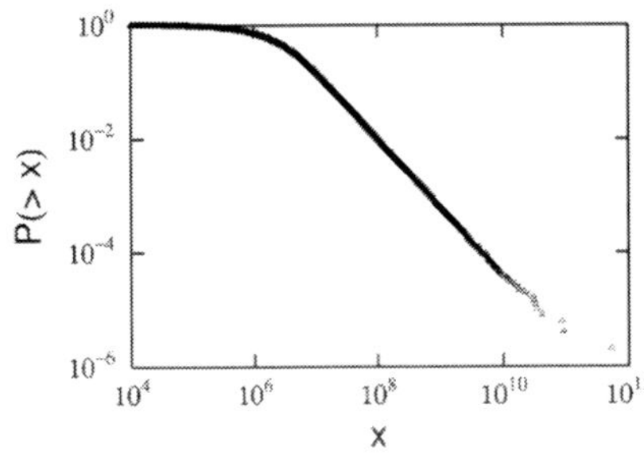

Figure 19. The resultant cumulative pdf $P(v)$.

5. Conclusion and future issues

In this paper, by using numerical simulations, we explain the difference between two types of the growth rate distributions. One is observed in both positive and negative data such as profits, the distribution of which can be approximated by linear functions on a log–log scale. The other is observed in non–negative data such as sales, the distribution of which has wide tails.

In the previous study [24], the simulation model generates only non–negative variables. At the same time, the growth rate distributions are adopted to be linear functions on a log–log scale. To resolve this discrepancy, in this study, the model has been modified to generate both positive and negative variables. The variables follow Detailed Balance and First Non–Gibrat's Law in the second, third and fourth quadrants, as observed in the first quadrant of the original model [24]. This is consistent with the results of analyses in temporal changes of sales data [9].

Using both positive and negative variables v, we have obtained non–negative variables x by assuming the change of x to be v. In the joint pdf of x, we have confirmed Detailed Balance. At the same time, we have found that the growth rate distributions have wide tails and Second Non–Gibrat's Law is observed in the middle scale range. As a matter of fact, Second Non–Gibrat's Law with respect to x is not related to First Non–Gibrat's Law with respect to v because Second Non–Gibrat's Law is also observed in the case that the growth rate distribution of v follows Gibrat's Law only ($\alpha = 0$). Second Non–Gibrat's Law is induced from the superposition of variables v. This result may be interpreted as the relation between stock and flow.

In this paper, we assume that there are similar joint pdfs in four quadrants of v. However, they are thought to be different from each other in an unstable economy [26, 27]. This can be controlled by the factor a in Eq. (11). As a next step, quantitative analyses of the simulation including the above situation are under investigation. The analytic derivation for the result in the simulation should be also examined. The reason will be clarified why the growth rate distributions of v are approximated by linear functions on a log–log scale. These researches might provide useful knowledge in the management of credit risk.

Acknowledgments

The authors are grateful to APFA7 & Tokyo Tech – Hitotsubashi Interdisciplinary Conference, where this work was completed and especially to Prof. H. Yoshikawa and Prof. T. Watanabe for useful comments. This work was supported in part by a Grant–in–Aid for Scientific Research (C) (No. 20510147) from the Ministry of Education, Culture, Sports, Science and Technology, Japan.

[1] Pareto V 1897 *Cours d'Economique Politique* (London: Macmillan)
[2] Kesten H 1973 *Acta Math.* **131** 207
[3] Levy M and Solomon S 1996 *Int. J. Mod. Phys.* C **7** 595
[4] Sornette D and Cont R 1997 *J. Phys.* I **7** 431
[5] Takayasu H, Sato A and Takayasu M 1997 *Phys. Rev. Lett.* B **79** 966
[6] Fujiwara Y, Souma W, Aoyama H, Kaizoji T and Aoki M 2003 *Physica* A **321** 598
[7] Fujiwara Y, Guilmi C D, Aoyama H, Gallegati M and Souma W 2004 *Physica* A **335** 197
[8] Gibrat R 1932 *Les inegalites economiques* (Paris: Sirey)
[9] Ishikawa A 2009 Power-Law and Log-Normal Distributions in Temporal Changes of Firm-Size Variables *Economics* **3** 2009-11
[10] TOKYO SHOKO RESEARCH, LTD., http://www.tsr-net.co.jp/
[11] Amaral L A N, Buldyrev S V, Havlin S, Leschhorn H, Maass P, Salinger M A, Stanley H E and Stanley M H R 1997 *J. Phys. I France* **7** 621
[12] Okuyama K, Takayasu M and Takayasu H 1999 *Physica* A **269** 125
[13] Matia K, Fu D, Buldyrev S V, Pammolli F, Riccaboni M and Stanley H E 2004 *Europhys. Lett.* **67** 498
[14] Gabaix X 2005 *The Granular Origins of Aggregate Fluctuations* (MIT and NBER Working Paper)
[15] Ishikawa A 2006 *Physica* A **363** 367
[16] Canning D, Amaral L A N, Lee Y, Meyer M and Stanley H E 1998 *Economics Lett.* **60** 335
[17] Badger W W 1980 *Mathematical Models as a Tool for the Social Science* (New York: Gordon and Breach) p 87
[18] Montroll E W and Shlesinger M F 1983 *J. Stat. Phys.* **32** 209
[19] Ishikawa A 2006 *Physica* A **367** 425
[20] Ishikawa A 2007 *Physica* A **383** 79
[21] Aoyama H 2004 *9th Ann. Workshop on Economic Heterogeneous Interacting Agents* (Kyoto)
[22] Aoyama H, Fujiwara Y and Souma W 2004 *The Physical Society of Japan* 2004 *Autumn Meeting* (Aomori)
[23] Takayasu H 2009 *APFA7& Tokyo Tech - Hitotsubashi Interdisciplinary Conf.* (Tokyo)
[24] Tomoyose M, Fujimoto S and Ishikawa A 2009 *Proc. of the YITP Workshop on Econophysics* (Kyoto) *Prog. Theor. Phys. Suppl.* **179** 114
[25] Mizuno T, Takayasu M and Takayasu H 2004 *Physica* A **332** 403
[26] Ishikawa A 2006 *Physica* A **371** 525
[27] Ishikawa A 2007 *Proc. of the YITP Workshop on Econophysics* (Kyoto) *Prog. Theor. Phys. Suppl.* **179** 103

A multiplicative stochastic process deriving the probability distribution in exact form

Shouji Fujimoto[1], **Atushi Ishikawa**[1] **and Masashi Tomoyose**[2]

[1]Kanazawa Gakuin University, 920-1392, Japan
[2]University of the Ryukyus, 903-0213, Japan

E-mail: fujimoto@kanazawa-gu.ac.jp

Abstract. A simple multiplicative stochastic process is proposed in order to clarify directly the origin of the power–law. In this model, the probability distribution of the stochastic variables is calculated in an exact form. The necessary conditions that the multiplicative stochastic process generates the positive–negative asymmetric power–law distribution are investigated. Finally, comparisons between the distribution of stochastic variables simulated numerically and the stable distribution are discussed.

1. Introduction

Power-law distributions are frequently observed in various kinds of phenomena not only in natural science but also in social science. For instance, distributions of assets, sales, profits, income of firms, the number of employees and personal income (denoted by x) follow the power–law in the large scale range. The cumulative probability greater than x is represented as

$$P_{>}(x) \propto x^{-\alpha}. \tag{1}$$

This is known as Pareto's Law [1], which has been well investigated by using various models. One of the useful models is a multiplicative stochastic process. This model is given by Langevin equation with multiplicative stochastic noises b_n and additive stochastic noises f_n:

$$x_{n+1} = b_n x_n + f_n. \tag{2}$$

The stochastic variables x_n are updated iteratively, and b_n, f_n and x_n are independent each other.

In the stochastic process, the conditions that the distribution of x follows the power–law are studied[2, 3]. And it is also shown that the power–law index α is estimated by the condition

$$\langle b_n^{\alpha} \rangle = 1. \tag{3}$$

Specifically, these conditions are shown by Kesten[2] using renewal theory, and are also shown by Takayasu–Sato–Takayasu[3] by comparing, in terms of the characteristic function, between the multiplicative stochastic process and the general power–law distribution. The mechanism that the multiplicative stochastic process generates the power–law distribution is investigated in Refs.[4, 5]. The main purpose of this study is to understand the mechanism by calculating the probability function of x_n directly.

For this aim, the authors introduce the multiplicative stochastic model, the noises of which are set to be simple. In this simple model, the probability function can be represented as an exact form analytically. We directly find that the analytic representation follows the power–law.

Under this simple multiplicative stochastic model, it is possible to investigate the power–low distribution which has the asymmetry of positive and negative. Because the data such as assets, sales and the number of employees take only positive values, the distribution follow the power–law only on the positive side. Therefore it is important to investigate the asymmetrical distribution without negative values. In this study, the necessary conditions that the multiplicative stochastic process generates the asymmetric distribution are first investigated. The general form of the stable distribution is considered especially with respect to the asymmetric distribution which has the power–law tail.

Moreover, numerical simulations are employed in order to extend the analytical solvable model. It is numerically confirmed that the distribution of stochastic variables agrees with the stable distribution under a certain condition.

2. Analytically solvable multiplicative stochastic process

In this section, the authors propose a model which has an analytic solution in order to clarify the origin of the power–law. Let us consider the simple case [1] that additive noises f_n are not stochastic but equal to some constant C in Eq.(2) as follows

$$x_{n+1} = b_n x_n + C, \tag{4}$$

where the probability function of multiplicative noises b_n are given by

$$q(b_n) = \begin{cases} 1-p & (b_n = 0) \\ p & (b_n = \lambda) \\ 0 & (\text{otherwise}) \end{cases} . \tag{5}$$

This is Bernoulli distribution multiplied by a scale factor λ. In this case, the power–law index α is estimated to be

$$\alpha = \log_\lambda p^{-1} \tag{6}$$

by using the necessary condition (3).

From Eq.(4), x_{n+1} is represented as

$$x_{n+1} = C(1 + b_n + b_n b_{n-1} + \cdots + b_n b_{n-1} \cdots b_2 b_1) + b_n b_{n-1} \cdots b_1 b_0 x_0. \tag{7}$$

In this model, by using Eq.(7), probabilities of x_{n+1} can be exactly calculated. The results are shown in Table 1. Here, $p(x_{n+1})$, $P(x_{n+1})$ and $P_>(x_{n+1})$ are a probability of x_{n+1}, a cumulative probability less than x_{n+1} and a cumulative probability greater than x_{n+1}, respectively. Here, for simplicity, we have set $x_0 = 0$. From Table 1, the cumulative probability is represented as

$$P_>\left(x_{n+1} = C\frac{\lambda^{i+1} - 1}{\lambda - 1}\right) = p^i \qquad (i = 0, \cdots, n). \tag{8}$$

Furthermore, to examine continuous behavior of the cumulative probability (8), let us define

$$x = C(\lambda^{i+1} - 1)/(\lambda - 1), \tag{9}$$
$$y = p^i. \tag{10}$$

[1] In Ref.[4], the simple case is considered that multiplicative noises b_n are constant and additive noises f_n are stochastic variables, the distribution of which is power–law.

Table 1. A probability of x_{n+1}, a cumulative probability less than x_{n+1} and a cumulative probability greater than x_{n+1}

x_{n+1}	$p(x_{n+1})$	$P(x_{n+1})$	$P_>(x_{n+1})$
C	$1-p$	$1-p$	1
$C(1+\lambda)$	$p(1-p)$	$1-p^2$	p
$C(1+\lambda+\lambda^2)$	$p^2(1-p)$	$1-p^3$	p^2
$\cdots$	$\cdots$	$\cdots$	$\cdots$
$C(1+\lambda+\lambda^2+\cdots+\lambda^{n-1})$	$p^{n-1}(1-p)$	$1-p^n$	p^{n-1}
$C(1+\lambda+\lambda^2+\cdots+\lambda^{n-1}+\lambda^n)$	p^n	1	p^n

The property of the distribution of x is quite different between in the case of $\lambda > 1$ and in the case of $\lambda < 1$. In the case of $\lambda > 1$, x can be approximated by $x \sim C\lambda^{i+1}/(\lambda - 1)$ in the limit $i \to \infty$. Then the cumulative probability (8) is expressed as the power–law:

$$y \sim Dx^{-\alpha}. \tag{11}$$

Here α is given by Eq.(6) and D is a constant value as follows

$$D = \left(C\frac{\lambda}{\lambda - 1}\right)^{\alpha}. \tag{12}$$

In the case of $\lambda < 1$, there is an upper bound of the distribution of x, $C/(1-\lambda)$. Figure 1 shows a numerical simulation result, the cumulative probability (8) and the continuous representation (11) in the case of $\lambda > 1$. The simulation result agrees with the cumulative probability (8) accurately. In this figure, as x is larger, the cumulative probability (8) is quickly closer to the asymptotic continuous power–law (11). Figure 2 shows those in the case of $\lambda < 1$. It is observed that the distribution of x is not power–law.

In addition, in this simple model, the average of x_{n+1} is obtained as follows:

$$\begin{aligned} \langle x_{n+1}\rangle &= C(1-p)\sum_{i=0}^{n-1} p^i S_i + Cp^n S_n \\ &= C\frac{1-(p\lambda)^{n+1}}{1-p\lambda}, \end{aligned} \tag{13}$$

where $S_i = (1+\lambda+\lambda^2+\cdots+\lambda^i)$. In the case of $p\lambda \geq 1$, this average diverges in the limit $n \to \infty$. [2] In the case of $\lambda > 1$, the condition $p\lambda \geq 1$ is rewritten as $1 \geq \log p^{-1}/\log\lambda$. This is identical with $\alpha \leq 1$ (see Eq.(6)). The divergence of the moment of order $m\,(\geq 2)$ is estimated by using a following inequality:

$$\begin{aligned} \langle x_{n+1}^m\rangle &= C^m(1-p)\sum_{i=0}^{n-1} p^i S_i^m + C^m p^n S_n^m \\ &> C^m\frac{1-(p\lambda^m)^{n+1}}{1-p\lambda^m}. \end{aligned} \tag{14}$$

At least in the case of $p\lambda^m \geq 1$, the moment of order m diverges in the limit $n \to \infty$. This condition is identical with $\alpha \leq m$ in the case of $\lambda > 1$.

[2] This model means St. Petersburg paradox in the case of $p = 1/2$, $\lambda = 2$ and $C = 1$.

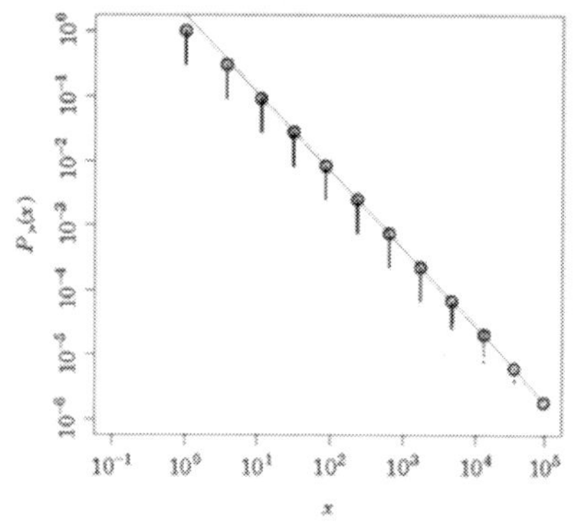

Figure 1. The cumulative probability greater than x_{n+1}: $P_>(x)$. Black dots, which seem to be vertical lines, are simulation results for $\lambda = e (> 1)$, $p = e^{-1.2}$ and $C = 1$. Circle marks are the cumulative probability (8) shown in Table 1. The gray straight line is the continuous power–law (11).

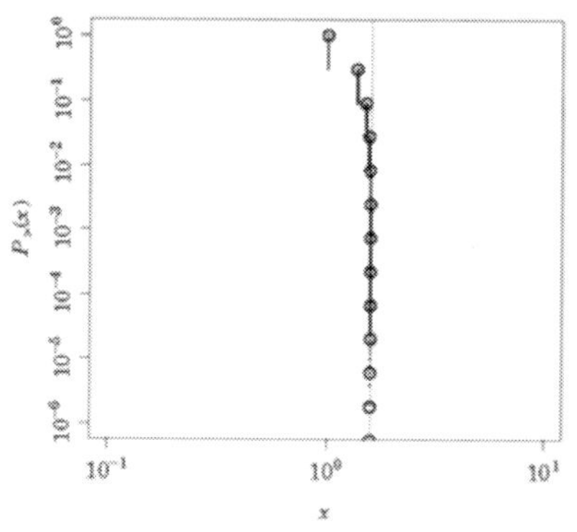

Figure 2. The cumulative probability greater than x_{n+1}: $P_>(x)$. Black dots, which seem to be vertical lines, are simulation results for $\lambda = e^{-1} (< 1)$, $p = e^{-1.2}$ and $C = 1$. Circle marks are the cumulative probability (8) shown in Table 1. The gray straight line is the upper bound of $x = C/(1-\lambda)$.

3. Effect of additive noises

In the previous section, in order to carry out the analytical calculation, additive noises and multiplicative noises are set to be constant and 0 or λ, respectively. In such a case, x follows the discrete distribution (8), the intervals of which are $C\lambda^i$ $(i = 1, \cdots, n)$. In this section, by taking various distributions of positive additive noises f_n (≥ 0), the effects of additive noises are examined. The purpose is to investigate how continuous power–law distributions are obtained. In the investigation, multiplicative noises b_n are given by the same distributions in the previous section (5) with $\lambda = e (> 1)$ and $p = e^{-1.2}$.

Firstly, let us consider the case that f_n follow two-values distributions. The probability function is given by

$$r(f_n) = \begin{cases} p_1 & (f_n = C_1) \\ p_2 & (f_n = C_2) \\ 0 & (\text{otherwise}) \end{cases}, \tag{15}$$

where $p_1 + p_2 = 1$. Although it is difficult to obtain an analytic representation of probabilities of x, the distribution of x can be investigated by a numerical simulation. Figure 3 shows the simulation result. The resulting distribution have been denser compared with the discrete distribution in the previous section (Fig. 1).

Secondarily, let us consider the case that f_n follow continuous uniform distributions. The probability density function is given by

$$r(f_n) = \begin{cases} \frac{1}{\lambda_f} & (0 \leq f_n \leq \lambda_f) \\ 0 & (\text{otherwise}) \end{cases}. \tag{16}$$

Figure 4 shows the simulation result, in which the distribution of x seems to be continuous. In the power–law range in Fig. 4, the distribution of x oscillates. This oscillation is attributed to the discontinuity of additive noises at $f_n = \lambda_f$.

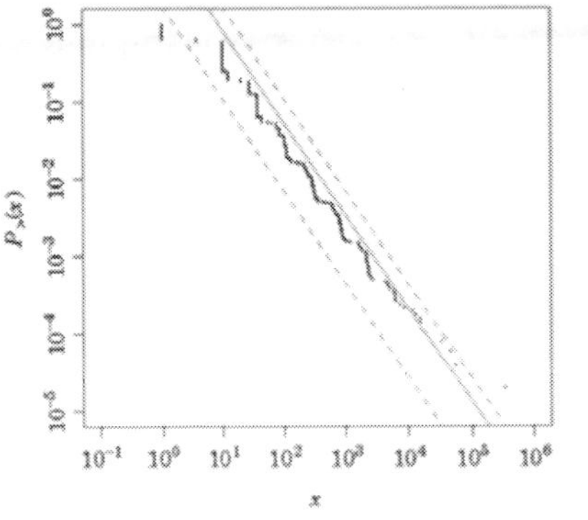

Figure 3. The simulation result in the case that additive noises follow two-values distributions (15). Black dots, which seem to be meandering curve, are simulation results for $C_1 = 1$, $C_2 = 10$ and $p_1 = p_2 = 0.5$. The gray solid straight line follows Eqs.(11)-(12), in which the parameter C is replaced by $\langle f_n \rangle$. The lower gray dashed line follows Eqs.(11)-(12), in which the parameter C is replaced by C_1. The upper gray dashed line follows Eqs.(11)-(12), in which the parameter C is replaced by C_2.

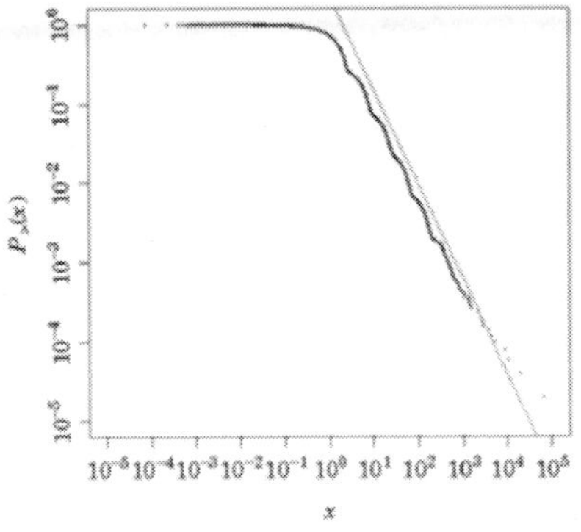

Figure 4. The simulation result in the case that additive noises follow continuous uniform distributions (16). Black dots, which seem to be meandering curve, are simulation results for $\lambda_f = e$. The gray straight line follows Eqs.(11)-(12), in which the parameter C is replaced by $\langle f_n \rangle$.

Thirdly, let us consider the case that f_n follow exponential distributions. The probability density function is given by

$$r(f_n) = \begin{cases} \frac{1}{\lambda_f} e^{-f_n/\lambda_f} & (0 \leq f_n) \\ 0 & (\text{otherwise}) \end{cases} . \tag{17}$$

Figure 5 shows the simulation result. The distribution of x have been smooth and continuous power–law.

Finally, let us consider the case that f_n follow Weibull distributions. The probability density function is given by

$$r(f_n) = \begin{cases} \frac{k_f}{\lambda_f} \left(\frac{f_n}{\lambda_f} \right)^{k_f - 1} e^{-(f_n/\lambda_f)^{k_f}} & (0 \leq f_n) \\ 0 & (\text{otherwise}) \end{cases} . \tag{18}$$

Figure 6 shows the simulation result. The distribution of x oscillates. Furthermore, it dose not seem to be continuous in the large x range. These reasons are attributed to a sharp peak of additive noises distributions. Weibull distribution with large degrees of freedom k_f has a sharp peak in a certain positive value. Because the sharp peak is similar to the δ–function, it is thought that the result is similar to the results observed in the case we take constant additive noises.

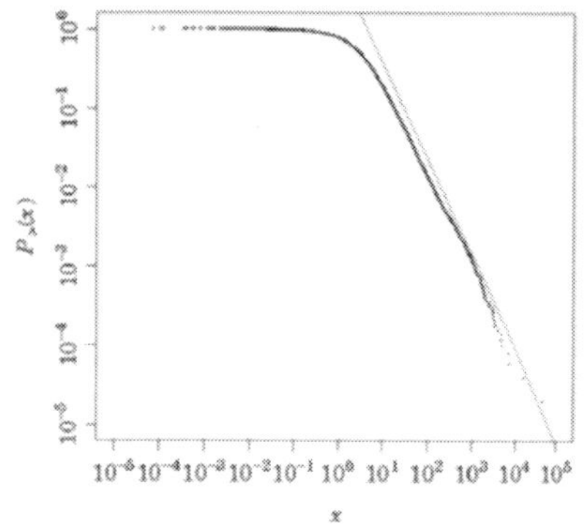

Figure 5. The simulation result in the case that additive noises follow exponential distributions (17). Black dots, which seem to be solid curve, are simulation results for $\lambda_f = e$. The gray straight line follows Eqs.(11)-(12), in which the parameter C is replaced by $\langle f_n \rangle$.

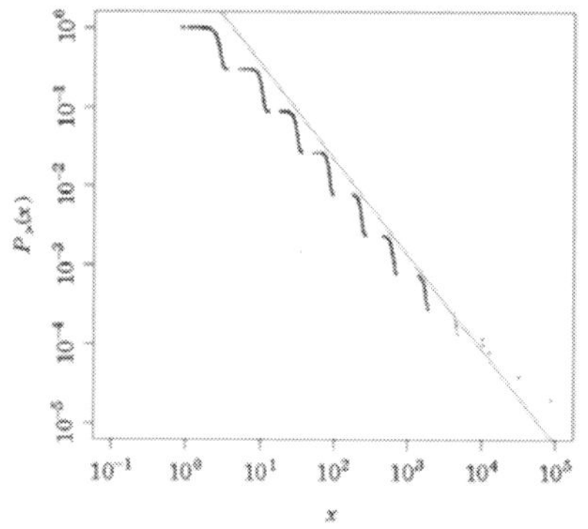

Figure 6. The simulation result in the case that additive noises follow Weibull distributions (18). Black dots, which seem to be meandering curve, are simulation results for $\lambda_f = e$ and $k_f = 10$. The gray straight line follows Eqs.(11)-(12), in which the parameter C is replaced by $\langle f_n \rangle$.

4. The asymmetric stable distribution

In the previous section, the multiplicative stochastic process which generates continuous power-law distributions is obtained by introducing continuous additive noises distributions. In this section, let us consider the necessary condition that the multiplicative stochastic process derives the stable distribution. The purpose of this section is to investigate whether the distribution in previous section satisfies the condition for the stable distribution.

A stable distribution does not change the function form by the convolution, and the characteristic function is given by

$$\phi(t) = \exp\left\{it\delta - \gamma|t|^{\alpha}\left[1 - i\beta \mathrm{sgn}(t)\Phi(\alpha)\right]\right\}, \tag{19}$$

where $\mathrm{sgn}(t)$ is the sign of t and $\Phi(\alpha)$ is given by

$$\Phi(\alpha) = \begin{cases} \tan(\alpha\pi/2) & \text{for} \quad \alpha \neq 1 \\ -(2/\pi)\log|t| & \text{for} \quad \alpha = 1 \end{cases}. \tag{20}$$

Here, $\alpha \in (0, 2]$ gives the power-law index of the tail of the distribution, $\beta \in [-1, 1]$, called the skewness parameter, is a measure of asymmetry, $\gamma \in [0, \infty)$ is a scale parameter and $\delta \in (-\infty, \infty)$ is a location parameter.

The necessary condition that the multiplicative stochastic process generates the stable distribution has been given by Takayasu–Sato–Takayasu (TST) [3]. Conditions $\langle f_n^2 \rangle > 0$ and $\langle b_n^2 \rangle > 1$ are obtained from the requirement that the second order moment of x diverges. The condition $\langle b_n^{\alpha} \rangle = 1$ is obtained from the comparison between the characteristic functions. Conditions $\langle \ln b_n \rangle > 0$ and $\langle b_n^2 \rangle > 1$ are obtained by restricting α to the range of $(0, 2)$ under the condition of $\langle b_n^{\alpha} \rangle = 1$. The condition $\langle b_n^{\alpha} \rangle = 1$ is extended to $\langle |b_n|^{\alpha} \rangle = 1$ for Gaussian multiplicative noises in Ref.[6]. These arguments were restricted to the case of $\beta = 0$ and

$\delta = 0$, which means that the distribution of x has positive–negative symmetry. The distribution obtained in the previous section takes only positive values and, therefore, the distribution has left-right asymmetry. In order to compare such the distribution with the stable distribution, it is necessary to examine whether these conditions given by TST can be applied to the case of $\beta \neq 0$ and $\delta \neq 0$.

The probability density functions of the distribution of x_n, b_n and f_n are expressed as $p(x_n)$, $q(b_n)$ and $r(f_n)$, respectively. The characteristic functions of x_n and f_n are expressed as $\phi_n(t)$ and $\psi(t)$, respectively. The characteristic function on the left hand side of Eq.(2) is

$$\phi_{n+1}(t) = \langle e^{itx_{n+1}} \rangle = \int_{-\infty}^{\infty} e^{itx_{n+1}} p(x_{n+1}) dx_{n+1}. \tag{21}$$

The characteristic function on the right hand side of Eq.(2) is

$$\begin{aligned} \langle e^{it(b_n x_n + f_n)} \rangle &= \int_{-\infty}^{\infty} \left\{ \int_{-\infty}^{\infty} e^{itb_n x_n} p(x_n) dx_n \right\} q(b_n) db_n \times \int_{-\infty}^{\infty} e^{itf_n} r(f_n) df_n \\ &= \int_{-\infty}^{\infty} \phi_n(b_n t) q(b_n) db_n \times \psi(t). \end{aligned} \tag{22}$$

Then the expression

$$\phi_{n+1}(t) = \int_{-\infty}^{\infty} \phi_n(b_n t) q(b_n) db_n \times \psi(t) \tag{23}$$

is obtained. In the limit $n \to \infty$, supposing x_{n+1} and x_n become the stable distributions with same parameters, these characteristic functions are written by $\phi(t)$ (19). Moreover, f_n is assumed to be the distribution, the characteristic function of which is expanded only by the integer order in terms of t. The expansion of each characteristic function in Eq.(23) around $t = 0$ is given by

$$\phi_{n+1}(t) = \phi(t) = 1 + it\delta - \gamma|t|^\alpha \left[1 - i\beta \mathrm{sgn}(t)\Phi(\alpha)\right] + \cdots, \tag{24}$$

$$\phi_n(b_n t) = \phi(b_n t) = 1 + ib_n t\delta - \gamma|b_n t|^\alpha \left[1 - i\beta \mathrm{sgn}(b_n t)\Phi(\alpha)\right] + \cdots, \tag{25}$$

$$\psi(t) = 1 + it\langle f_n \rangle + \cdots. \tag{26}$$

By substituting Eqs.(24)-(26) into Eq.(23) and comparing coefficients in terms of t, necessary conditions that the distribution of x is stable are obtained.

The location parameter δ is obtained by comparing the coefficient of t in both sides of Eq.(23) as follows

$$\delta = \int_{-\infty}^{\infty} b_n \delta\, q(b_n) db_n + \langle f_n \rangle \tag{27}$$

$$= \frac{\langle f_n \rangle}{1 - \langle b_n \rangle}. \tag{28}$$

From t^α order term in Eq.(23), the condition

$$\gamma|t|^\alpha \left[1 - i\beta \mathrm{sgn}(t)\Phi(\alpha)\right] = \int_{-\infty}^{\infty} \gamma|b_n t|^\alpha \left[1 - i\beta \mathrm{sgn}(b_n t)\Phi(\alpha)\right] q(b_n) db_n \tag{29}$$

is obtained. In the case of $t > 0$, two conditions

$$\gamma = \gamma \left[\int_{-\infty}^{0} (-b_n)^\alpha q(b_n) db_n + \int_{0}^{\infty} b_n^\alpha q(b_n) db_n \right], \tag{30}$$

$$i\gamma\beta\Phi(\alpha) = i\gamma\beta\Phi(\alpha) \left[-\int_{-\infty}^{0} (-b_n)^\alpha q(b_n) db_n + \int_{0}^{\infty} b_n^\alpha q(b_n) db_n \right] \tag{31}$$

are derived from the real and imaginary parts of Eq.(29). [3] In order to satisfy both Eqs.(30)-(31), at least one of the following four conditions is necessary:

- $\gamma = 0$ (x is constant δ)
- $\beta = 0$ (the distribution of x is symmetric)
- $\Phi(\alpha) = 0$ ($\alpha = 2$)
- $q(b_n) = 0$ for $b_n < 0$ (the distribution of b_n is positive)

In the case of $\gamma \neq 0$(x is not constant), $\beta = 0$(the distribution of x has positive–negative symmetry) and $\alpha \neq 2$(the distribution of x is not the normal distribution), it is allowed that b_n take negative values, and the power–law index α can be estimated by the condition $\langle |b_n|^\alpha \rangle = 1$ [6] by Eq.(30). In the case of $\gamma \neq 0$, $\beta \neq 0$(the distribution of x dose not have positive–negative symmetry) and $\alpha \neq 2$, it is not allowed that b_n take negative values, and the power–law index α can be estimated by the condition $\langle b_n^\alpha \rangle = 1$ [3] by Eq.(30). These arguments are valid even if the distribution of b_n is discrete.

From these results, it is considered that the negative multiplicative noises b_n have an effect which makes the distribution of x symmetrical. This is consistent with the analytic form of the distribution of x in section 2. Let us take $\lambda < -1$ in the Table 1. In such a case, the sign of x_{n+1} takes the value of positive and negative alternately. Therefore the distribution of x_{n+1} has the positive–negative symmetry.

In the previous section, the distribution of x does not have positive–negative symmetry and b_n take positive values. In such the case, the condition $\langle b_n^\alpha \rangle = 1$ should be taken. Since the conditions obtained in this section are the necessary conditions, the resulting distributions are not always stable even if these conditions are satisfied. This point will be investigated in the following section.

5. Comparison between the simulation result and the stable distribution

The multiplicative stochastic process, which can express the probability of stochastic variables exactly, is introduced in section 2. The model is extended to generate the continuous distribution in section 3. The necessary conditions for the asymmetric stable distribution are derived in section 4. In this section, it will be confirmed that the distribution generated by the multiplicative stochastic process is the stable distribution. In this discussion, Eq.(5) and Eq.(17) are adopted for the distribution of b_n and f_n, respectively.

The stable distribution is characterized by four parameters α, β, γ and δ in the characteristic function (19). It is necessary to identify the parameters in order to compare the simulation result with the stable distribution. The parameter α is identified by Eq.(6). In the case of $\alpha \geq 1$, the parameter δ is identified by Eq.(28). In order to decide other parameters β, γ and δ(for $\alpha < 1$), we use universal properties known about the stable distribution [7, 8].

Asymptotic behavior of the cumulative distribution of the stable distribution is given by

$$P_>(x) \sim \gamma^\alpha c_\alpha (1+\beta) x^{-\alpha} \tag{32}$$

as $x \to \infty$ in the case of $-1 < \beta \leq 1$, where $c_\alpha = \sin(\pi\alpha/2)\Gamma(\alpha)/\pi$ [7, 8]. Asymptotic behavior in the opposite side of the distribution of x is given by

$$P_>(-x) \sim \gamma^\alpha c_\alpha (1-\beta)(-x)^{-\alpha} \tag{33}$$

as $x \to -\infty$ in the case of $-1 \leq \beta < 1$ [7, 8]. For $\beta = \pm 1$, the stable distribution has one power–law tail. For $\beta \neq \pm 1$, the stable distribution has power–law tails in both sides. Since

[3] In the case of $t < 0$, the same conditions are derived from the coefficient of $(-t)^\alpha$.

the multiplicative stochastic process considered in this study has the power–law tail only in a positive side, we take $\beta = 1$. The parameter γ is fixed by Eq.(6), $\beta = 1$ and Eq.(32).

According to Ref.[7, 8], the support of the stable distribution is

$$\begin{array}{ll} x \in [\delta - \gamma \tan(\pi\alpha/2), \infty) & \alpha < 1 \text{ and } \beta = 1 \\ x \in (-\infty, \delta + \gamma \tan(\pi\alpha/2)] & \alpha < 1 \text{ and } \beta = -1 \ . \\ x \in (-\infty, \infty) & \text{otherwise} \end{array} \quad (34)$$

In the case of $\alpha < 1$ and $\beta = 1$, the support of the stable distribution is $x \in [\delta - \gamma \tan(\pi\alpha/2), \infty)$. Since the support of the distribution of the simulation result is $x \in [0, \infty)$, the parameter δ is identified by $\delta = \gamma \tan(\pi\alpha/2)$.

Firstly, let us compare a simulation result with the stable distribution in the case of $0 < \alpha < 1$. In order to set $\alpha = 0.5$, the parameters of Eq.(5) are adopted for $\lambda = e$ and $p = e^{-0.5}$. The parameter of Eq.(17) is adopted for $\lambda_f = 1$. Figure 7 shows the comparison between the cumulative distribution of the simulation result and the stable distribution in the case of $\alpha = 0.5$. Black points are the simulation results. The gray solid curve is the stable distribution. [4] The simulation result seems to agree with the stable distribution in terms of the cumulative distribution.

Figure 8 shows the same kind of the comparison by using of the probability density. Black points are the simulation results. The gray solid curve is the stable distribution. The gray dashed curve is the distribution of additive noises f_n. The simulation result seems to agree with the stable distribution in the range of $10^{-1} < x$. However, the simulation data do not agree with the stable distribution in the range of $x < 10^{-1}$. This reason is considered that the influence of additive noises becomes considerable in the small x. In the small x range, the probability density function of x becomes comparable with the probability density function of additive noises.

Secondly, let us compare a simulation result with the stable distribution in the case of $1 < \alpha < 2$. In order to set $\alpha = 1.5$, the parameters of Eq.(5) are adopted for $\lambda = e$ and $p = e^{-1.5}$. Moreover, in order to set $\delta = 1$, the parameter of Eq.(17) are adopted for $\lambda_f = 1/(1 - \langle b_n \rangle) = 1/(1 - e^{-0.5})$. The average value measured by simulation data is $\langle x \rangle = 0.9712$. This average almost agrees with δ. Figure 9 shows the comparison between the cumulative distribution of the simulation result and the stable distribution in the case of $\alpha = 1.5$. Black points are the simulation results. The gray solid curve is the stable distribution. [5] The simulation result seems to agree with the stable distribution only for $x > 10^1$ in the range of power–law. On the one hand, the support of the stable distribution $\alpha = 1.5$ and $\beta = 1.0$ is $x \in (-\infty, \infty)$. On the other hand, the support of the distribution of the simulation result is $x \in [0, \infty)$. Therefore it is hard to consider two distributions to be identified with each other. Figure 10 shows the same kind of the comparison by using the probability density function.

Our model does not completely reproduce the stable distribution. In the case of $0 < \alpha < 1$, the simulation result almost agrees with the stable distribution excluding the range of small x. In the case of $1 < \alpha < 2$, the simulation result dose not agree with the stable distribution, because the support of the distributions is quite different from the stable distribution.

[4] The parameter γ is estimated by fitting the cumulative distribution (32). The fitting range is taken from 99% to 99.9% of the distribution of x obtained by the simulation. As a result, the value of γ is estimated to be 1.6246 ± 0.0065. In the case of $\alpha = 0.5$, δ is the same value as γ, because $\delta = \gamma \tan(\pi\alpha/2) = \gamma$. For the parameters of the stable distribution, $\alpha = 0.5$, $\beta = 1.0$, $\gamma = 1.6246$ and $\delta = 1.6246$ are adopted. A program STABLE[9] is used to plot the stable distribution.

[5] The parameter γ is estimated by fitting the cumulative distribution (32). The fitting range is taken from 99% to 99.9% of the distribution of x obtained by the simulation. As a result, the value of γ is estimated to be $\gamma = 0.74980 \pm 0.00053$. For the parameter of the stable distribution, $\alpha = 1.5$, $\beta = 1.0$, $\gamma = 0.7480$ and $\delta = 1.0$ are adopted.

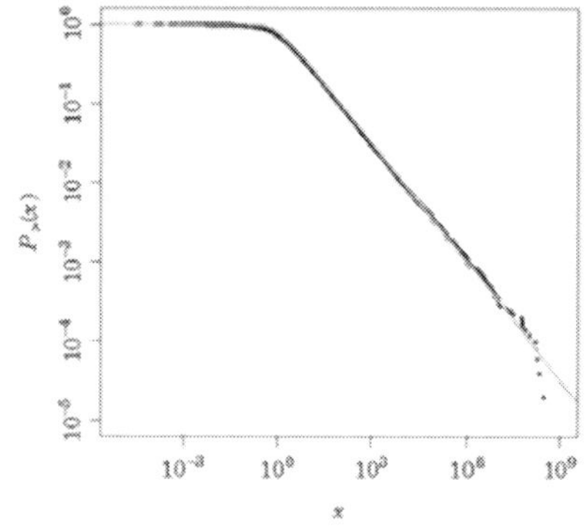

Figure 7. The cumulative probability greater than x: $P_>(x)$. The simulation result in the case of $\alpha = 0.5$ is shown. Black dots, which seem to be solid curve, are simulation results for $\lambda = e$, $p = e^{-0.5}$ and $\lambda_f = 1$. A gray curve is the stable distribution, the parameters of which are $\alpha = 0.5$, $\beta = 1.0$, $\gamma = 1.6246$ and $\delta = 1.6246$.

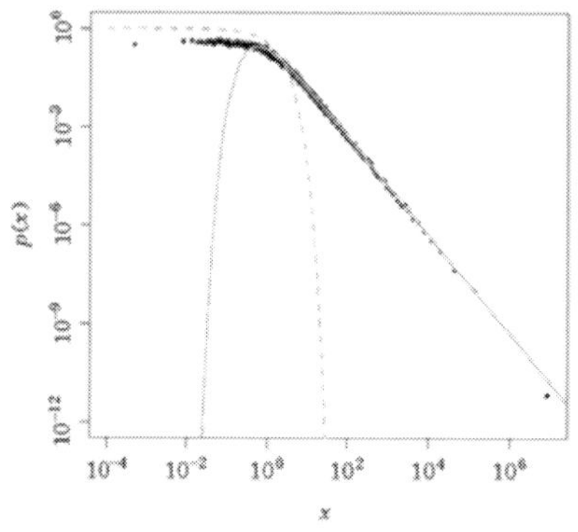

Figure 8. The probability density function: $p(x)$. The simulation result in the case of $\alpha = 0.5$ is shown. Black dots, which seem to be solid curve, are simulation results for $\lambda = e$, $p = e^{-0.5}$ and $\lambda_f = 1$. A solid gray curve is the stable distribution, the parameters of which are $\alpha = 0.5$, $\beta = 1.0$, $\gamma = 1.6246$ and $\delta = 1.6246$. A dashed gray curve is additive noises f_n (17).

6. Conclusion

The purpose of this study is to investigate the emergence of the power–law by the multiplicative stochastic process.

In section 2, the authors propose a model which derives the probability distribution of stochastic variables in exact form to clarify the origin of the power–law. In order to carry out an analytical calculation, additive noises and multiplicative noises are set to be constant and 0 or λ, respectively. The property of the distribution of x is quite different between in the case of $\lambda > 1$ and in the case of $\lambda < 1$. In the case of $\lambda > 1$, we find that the distribution is power–law. In the case of $\lambda < 1$, the distribution is not power–law.

In section 3, effects of additive noises are examined by numerical simulations. By taking various distributions of positive additive noises f_n (≥ 0), the continuous power–law distribution is obtained.

In section 4, the necessary conditions that the multiplicative stochastic process generates the positive–negative asymmetric stable distribution are first shown. The most important condition for the asymmetric distribution is that multiplicative noises take only positive values.

In section 5, the comparison between the distribution obtained by the numerical simulation and the stable distribution is examined. In the case of the power–law index $\alpha < 1$, the simulation result almost agrees with the stable distribution.

In this paper, the roles of b_n are clarified. The multiplicative noises of $b_n > 1$ extend the tail of the power–law distribution. The noises $b_n < 1$ limit the distribution finitely. The negative noises b_n make the distribution positive–negative symmetric. Let us consider that x_n are data of velocity. The noises $b_n > 1$ have the effect corresponding to acceleration. The noises $b_n < 1$ have

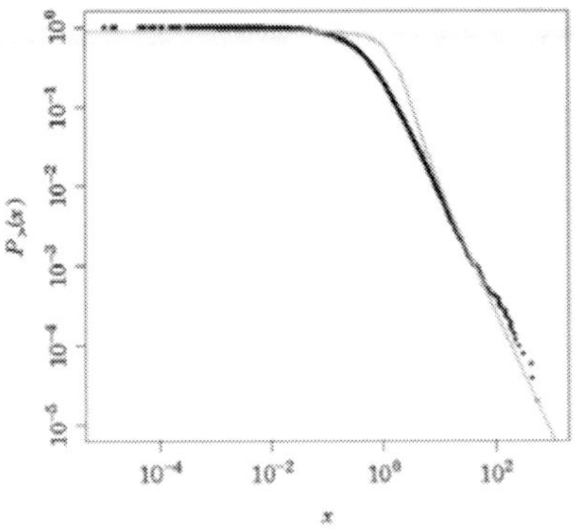

Figure 9. The cumulative probability greater than x: $P_>(x)$. The simulation result in the case of $\alpha = 1.5$ is shown. Black dots, which seem to be solid curve, are simulation results for $\lambda = e$, $p = e^{-1.5}$ and $\lambda_f = 1/(1 - e^{-0.5})$. A gray curve is the stable distribution, the parameters of which are $\alpha = 1.5$, $\beta = 1.0$, $\gamma = 0.7480$ and $\delta = 1.0$.

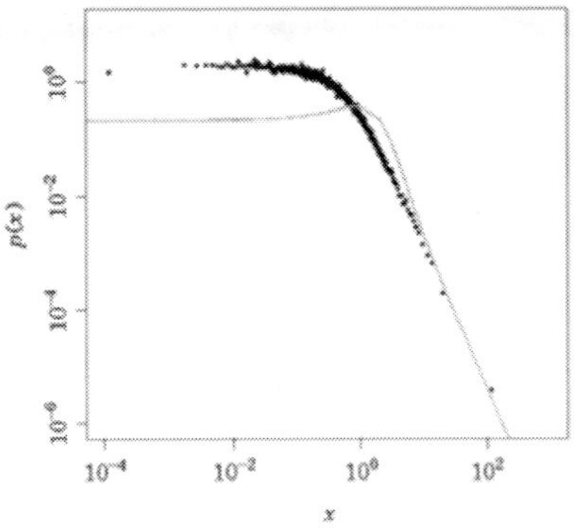

Figure 10. The probability density function: $p(x)$. The simulation result in the case of $\alpha = 1.5$ is shown. Black dots, which seem to be solid curve, are simulation results for $\lambda = e$, $p = e^{-1.5}$ and $\lambda_f = 1/(1-e^{-0.5})$. A gray curve is the stable distribution, the parameter of which are $\alpha = 1.5$, $\beta = 1.0$, $\gamma = 0.7480$ and $\delta = 1.0$.

the effect corresponding to deceleration. The negative noises b_n have the effect corresponding to reflection.

In the case of the general multiplicative stochastic process, various distributions are taken as the multiplicative noises and the additive noises. Although in such a complicated case, it is considered that the similar mechanism observed in this study produces the power–law distribution. For various applications of the multiplicative stochastic process, the authors believe that the knowledge obtained in this study will be useful significantly.

Acknowledgements
The authors are grateful to APFA7 & Tokyo Tech – Hitotsubashi Interdisciplinary Conference, where this work was completed. The authors are also grateful to J. P. Nolan who put the program STABLE on the website. This work was supported in part by a Grant–in–Aid for Scientific Research (C) (No. 20510147) from the Ministry of Education, Culture, Sports, Science and Technology, Japan.

References

[1] Pareto V 1897 *Cours d'Economique Politique* (London: Macmillan).
[2] Kesten H 1973 *Acta Math.* **131** 207.
[3] Takayasu H, Sato A and Takayasu M 1997 *Phys. Rev. Lett.* **79** 966.
[4] Sornette D 2004 *Critical phenomena in natural sciences* (Springer) p 374
[5] Sornette D 1998 *Phys. Rev.* E **57** 4811
[6] Sato A, Takayasu H and Sawada Y 2000 *Phys. Rev.* E **61** 1081.
[7] Samorodnitsky G and Taqqu M S 1994 *Stable non-Gaussian random processes* (CRC Press) p 16
[8] Fofack H and Nolan J P 1999 *Extremes* **2:1** 39.
[9] The program STABLE is available from J. P. Nolan's website: academic2.american.edu/~jpnolan

Non-Gibrat's Law and the size dependence of growth rate distributions

Masashi Tomoyose[1], **Shouji Fujimoto**[2] **and Atushi Ishikawa**[2]
[1]University of the Ryukyus, 903-0213, Japan
[2]Kanazawa Gakuin University, 920-1392, Japan

E-mail: `mtomo@weiss.phys.u-ryukyu.ac.jp`

Abstract. The authors study growth rate distributions of two types of variables by using numerical simulations. The first type data v exhibit both positive and negative values, and the second type data x only take positive values. The simulation model is defined by the Langevin equation for variables v, and variables x are obtained from v. Two kinds of Pareto indices, and the standard deviation σ of the growth rate distribution of x are acquired. Pareto indices for v and x are equal. Also, σ exhibits scaling behavior with a scaling exponent $\varepsilon < 0$. The scaling exponent ε depends on the Pareto index μ.

1. Introduction

There are many studies of distributions of sizes and growth rates of firms. In those studies, several empirical laws have been described. One of these is Pareto's Law [1]:

$$P(v) \propto v^{-(\mu+1)} \quad \text{for} \quad v > v_{\rm th}. \tag{1}$$

Here, v denotes a company's size in terms of its assets, sales, profits, the number of employees or some similar measure (see for instance [2]), and $P(v)$ is the probability density function (pdf)[1]. μ is called the Pareto index and $v_{\rm th}$ is a certain threshold.

Another empirical law is Gibrat's Law concerning growth rate distributions [3]. The growth rate R is defined as $R = v_2/v_1$, where v_1 and v_2 are two successive measures of a firm's size. Gibrat's Law indicates that the conditional pdf $Q(R|v_1)$ is independent of the past firm's size v_1

$$Q(R|v_1) = Q(R) \tag{2}$$

over a large size range [4]. In Ref. [4], size is variously defined as personal income in Japan (the former of the two references), as total assets or sales in France, or as the number of employees in the UK (the later reference).

The above two laws are valid over a large size range and are related to each other[2]. Fujiwara et al. [4] show that Pareto's Law is derived from Gibrat's Law and the Law of Detailed Balance.

[1] Analysis of data for Japanese listed companies reveals that the distribution of market capitalization is predominantly log-normal. This data is peculiar, in that the distribution of sales is also predominantly log-normal. The reason is that listed companies produce highly restricted sales data. Their data is, however, exhaustive with respect to market capitalization.

[2] There are several studies which show that Gibrat's Law leads to Pareto's Law under certain conditions (for instance [5]).

Detailed Balance is the time-reversal symmetry ($v_1 \leftrightarrow v_2$) observed in a stable economy

$$P_{12}(v_1, v_2) = P_{12}(v_2, v_1), \tag{3}$$

where $P_{12}(v_1, v_2)$ is the joint pdf.

The study of Pareto's Law over a large size range is important, because this encompasses a large percent of the overall measurements of assets, sales, profits, or number of employees. At the same time, the study of the distribution in the medium size range is also important, since the majority of companies are medium-sized. It is considered that the log-normal distribution agrees well with the data in this size range. The log-normal distribution is not, however, derived from Gibrat's Law under conditions of Detailed Balance. In fact, it is reported that Gibrat's Law is not relevant empirically (for instance [6, 7, 8, 9, 10, 11]), i.e. that $Q(R|v_1)$ does depend on the past size v_1. In Ref. [10], it is reported that the dependence on the past size is determined by the condition that $Q(R|v_1)$ is expressed as an exponential tent-shaped function:

$$Q(R|v_1) = \text{Const.}\ R^{\mp t_\pm(v_1)-1} \qquad \text{for} \quad R \gtrless 1. \tag{4}$$

The deduced past size dependence is as follows:

$$t_\pm(v_1) = t_\pm(v_{\rm th}) \pm \alpha\ \ln \frac{v_1}{v_{\rm th}}, \tag{5}$$

where $\alpha = 0$ for $v_1 > v_{\rm th}$ and $\alpha \neq 0$ for $v_1 < v_{\rm th}$. In addition, the pdf $P(v)$ is also deduced to be the log-normal distribution

$$P(v) = C\ v^{-(\mu+1)} e^{-\alpha \ln^2 \frac{v}{v_{\rm th}}} \qquad \text{for} \qquad v < v_{\rm th} \tag{6}$$

in the medium size range [9]. Note that Detailed Balance is a requirement in the above derivation.

The assumed $Q(R|v_1)$ is a linear function on a log-log scale as shown in Fig. 1. Moreover, according to Eq. (5), the probability of positive growth decreases and the probability of negative growth increases symmetrically as the classification of v increases in the medium size range. Profit data for Japanese firms exhibits such behavior [10]. We call Eqs. (4) and (5) the First Non-Gibrat's Law.

There are, however, some growth rate distributions which cannot be approximated by Eqs. (4) and (5) [6, 8, 11]. These growth rate distributions exhibit a fat tail behavior as shown in Fig. 2. Moreover, the probability of both positive and negative growth rates decreases simultaneously as the classification of v_1 increases, as shown in Fig. 2. We call this behavior the Second Non-Gibrat's Law.

Figure 1. The growth rate pdf for profits of firms on logarithmic scales. First Non-Gibrat's Law.

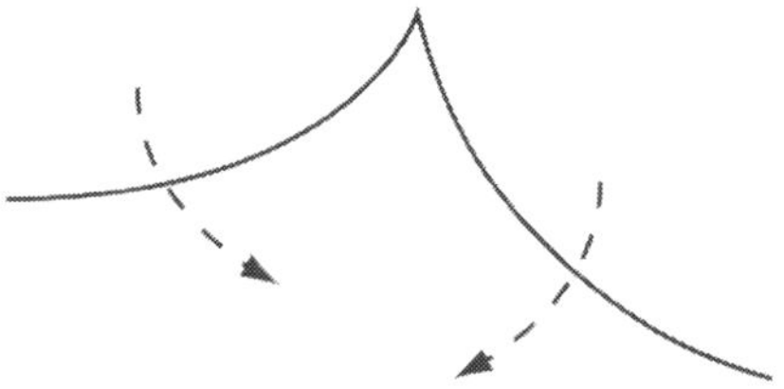

Figure 2. The growth rate pdf for sales or assets of firms on logarithmic scales. Second Non-Gibrat's Law.

One of the authors (A.I.) proposes an assumption to account for the difference between the two Non-Gibrat's Laws [12]. Then the economic variables may be classified into two types. One type is defined by using a kind of subtraction, whereas the other type does not involve any subtraction in its definition. Also it is assumed that the first type of data obeys the First Non-Gibrat's Law and the second type of data obeys the Second Non-Gibrat's Law. Moreover, first type data are obtained as the temporal difference between two second type data. For example, profits are data of the first type, and sales and assets are data of the second type. By definition, the first type data exhibits both positive and negative values. Regarding a scatter plot of v_1 vs v_2, the data exist not only in the first quadrant, but also in all four quadrants in the v_1-v_2 plane, where v_1 and v_2 express two successive firm sizes. In this study, variables v represent first type data, and x represent second type data.

In Ref. [13], we verify that the Takayasu-Sato-Takayasu (TST) model [14] satisfies not only Pareto's Law but also Detailed Balance under the First Gibrat's Law, by using numerical simulations. In the model, multiplicative stochastic noise obeying the First Non-Gibrat's Law has been employed. However, the subjects of the model were only positive data because both the multiplicative and the additive noise have been defined as positive. To avoid this discrepancy, an extension of the model is needed.

An extended model has been proposed in Ref. [15]. In this model, the variables v exhibit both positive and negative values. Positive variables x are then obtained from v. The growth rate distributions of x are also obtained. It is verified that the First and Second Non-Gibrat's Laws are valid for the growth rate distributions of v and x respectively.

In this study, the dependence of the obtained Second Non-Gibrat's Law on the past size x_1 is investigated by the use of numerical simulations. This study provides a clue to a quantitative analysis because it is important to determine what the Second Non-Gibrat's Law depends on. It can also serve as a clue to an analytic understanding.

2. Simulation model and results

We begin with a brief review of our model. First, the positive and negative stochastic variables v are generated by using the Langevin equation [16]:

$$v(t+1) = a(t) \cdot b(v(t), t) \cdot v(t) + f(t), \tag{7}$$

where $a = \pm 1$ with equal probability, b is a non-negative multiplicative stochastic noise and f is an additive stochastic noise. A normal distribution $N(0, s^2)$ is employed for the additive noise f. Eqs. (4) and (5) are utilized as the distribution of b. To avoid excessive changes of $t_\pm(v)$, $v_{\rm th2}$ is introduced, and the parameter α is chosen as $\alpha = 0$ for $v > v_{\rm th}$, $\alpha \neq 0$ for $v_{\rm th2} < v < v_{\rm th}$ and $\alpha = 0$ again for $v < v_{\rm th2}$ (Fig. 3). We also introduce the threshold $v_{\rm min}$, above which Detailed Balance is observed. Non-Gibrat's Law is observed in the range $\max(v_{\rm min}, v_{\rm th2}) < v < v_{\rm th}$. We call the combination of Gibrat's Law in the large scale range and Non-Gibrat's Law in the medium scale range as Extended-Gibrat's Law [9, 15].

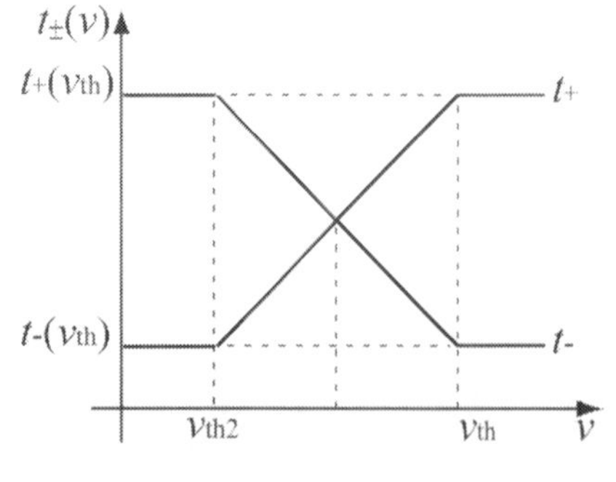

Figure 3. The shape of continuous functions $t_\pm(v)$. The horizontal axis is logarithmic scale.

Next, we define positive variables x by using both positive and negative variables v. It is proposed that v is provided by the temporal change of x:

$$v(t) = x(t+1) - x(t), \tag{8}$$

namely that the evolution of x can be expressed as

$$x(t+1) = \begin{cases} x(t) + v(t) & \text{if} \quad x(t) + v(t) > 0, \\ x(t) & \text{if} \quad x(t) + v(t) < 0. \end{cases} \tag{9}$$

Here, we restrict x to be positive. This restriction is relevant for $x \sim v$, but does not contribute for $x >> v$. The restriction does not affect results because variables x increase in general as the iteration proceeds. In the simulation, we do not add v less than $v_{\rm min}$ to x because Detailed Balance does not hold in the range below $v_{\rm min}$. We imposed this to avoid the effects of such data.

Let us describe the simulation results [15]. We first state a few results needed in the next section. Extended Gibrat's Law and Detailed Balance are confirmed. It is observed that the pdfs $P(v)$ and $P(x)$ follow Pareto's Law over a large scale range, for which the Pareto indices μ_v and μ_x are estimated.

The conditional pdfs $q(r|x_1)$ of the log growth rate $r = \log_{10} x_2/x_1$ are shown in Fig. 4. The pdf $q(r|x_1)$ is related to the pdf $Q(R|x_1)$ by $\log_{10} q(r|x_1) = \log_{10} Q(R|x_1) + r + \log_{10}(\ln 10)$. It is found that the Second Non-Gibrat's Law for variables x is valid in the medium scale range. The probability of positive and negative growth rate decrease simultaneously as the size of x_1 increases. The pdfs also exhibit a fat tail behavior. In the simulation, we employ the following parameters: $t_+(v_{\rm th}) = 2.6, t_-(v_{\rm th}) = 1.6, v_{\rm th} = 10^6, s = 20$ and $\alpha = 0$. The total number of data is 500,000.

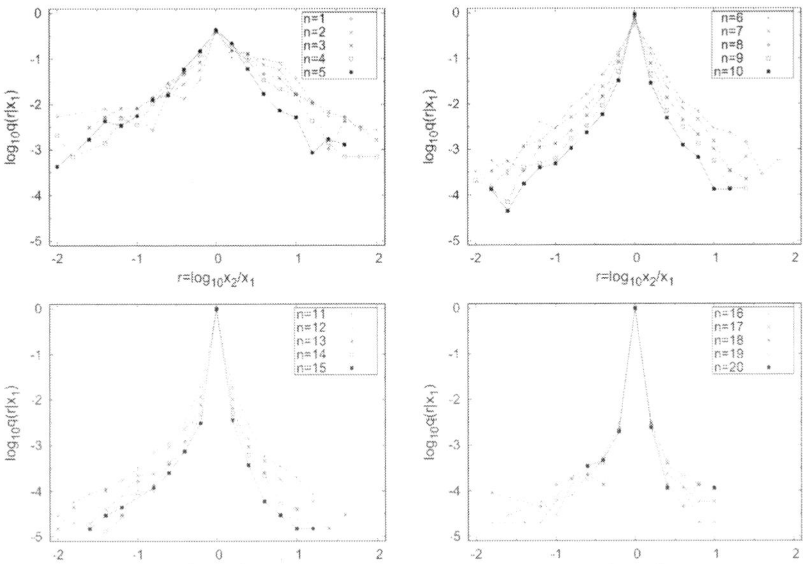

Figure 4. Conditional pdfs $q(r|x_1)$ of the log growth rate $r = \log_{10} x_2/x_1$. The data are classified into twenty bins of the past variables with equal magnitude in logarithmic scale, $x \in [10^{1+0.5(n-1)}, 10^{1+0.5n}]$ $(n = 1, 2, \ldots, 20)$.

3. Analysis of simulation results

We present several relations between the distribution of v and the distribution of x. First, let us compare μ_v and μ_x. As shown in Fig. 5, μ_x are equal to μ_v as follows:

$$\mu_x = \mu_v \equiv \mu \qquad \text{for } 0 < \mu < 2. \tag{10}$$

Next, let us consider the dependence of growth rate distributions on the past size x_1. As mentioned in section 1, the growth rate distribution of x obeys the Second Non-Gibrat's Law in the medium size range [15]. Namely, the width of the growth rate distribution of x decreases as the past size x_1 increases. To clarify this, we examine the standard deviation σ of the growth rate distribution of x.

Now we discuss the following two cases: (a) the growth rate distribution of v obeys only Gibrat's Law; (b) the growth rate distribution of v obeys the Extended Gibrat's Law. The result of case (a) is shown in Fig. 6. The dependence of σ on the past size is obtained as

$$\log_{10}\sigma = \text{Const.} + \varepsilon n, \tag{11}$$

$$\text{where} \qquad \varepsilon = -0.05\mu - 0.055 \qquad \text{for} \quad 1.0 < \mu < 1.5. \tag{12}$$

Here, n is a size parameter of x: $x \in [10^{1+0.5(n-1)}, 10^{1+0.5n}]$. So Eq. (11) can be expressed by

$$\sigma \propto x^{2\varepsilon}. \tag{13}$$

The standard deviation σ decreases as the size of x increases because ε is negative. This result is consistent with earlier works [6, 8, 11]. Furthermore, Eq. (12) indicates that the scaling exponent ε depends on μ (Fig. 7).

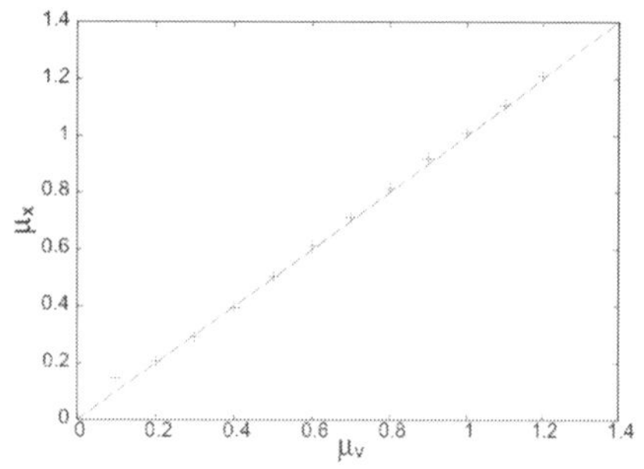

Figure 5. Comparison of μ_v and μ_x.

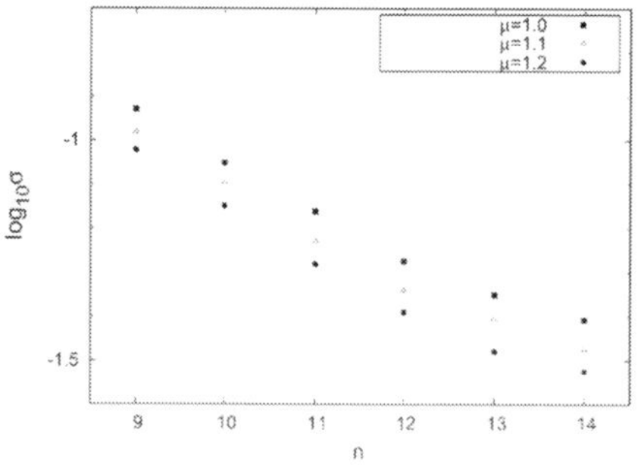

Figure 6. The dependence of σ on the past size x_1.

Let us consider the result of case (b). In case (b), the growth rate distribution of v obeys the First Non-Gibrat's Law, and the pdf of v forms the log-normal distribution (6) in the medium scale range [9]. The relation between ε and α is shown in Fig. 8. This dependence is obtained as

$$\varepsilon(\alpha) = 0.8\alpha + \varepsilon(\alpha = 0). \tag{14}$$

Here, we restrict ourselves to the case $\alpha < 0.1$. This restriction makes it possible to keep a certain width for the range in which the First Non-Gibrat's Law holds with respect to v. For the same reason, it is required that v_{th2} is less than or equal to v_{min}.

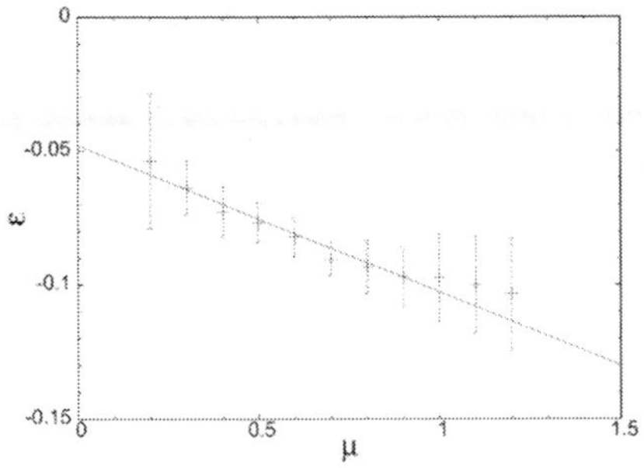

Figure 7. The dependence of ε on μ.

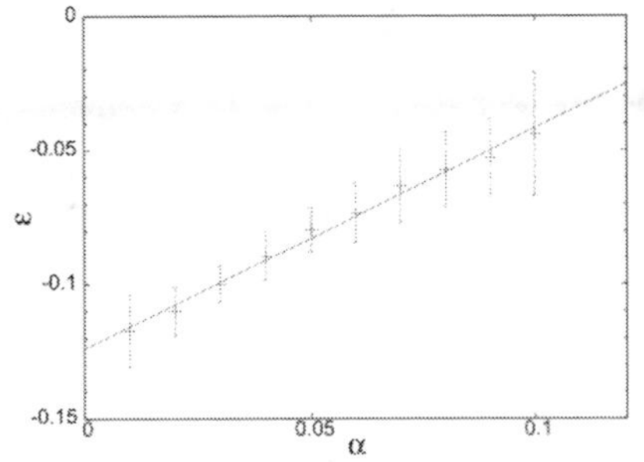

Figure 8. The dependence of ε on α.

4. Conclusion

Using numerical simulations, we study two types of variables and their growth rate distributions. The first type data v exhibit both positive and negative values, and the second type data x only take positive values. Moreover, variables v are taken as the temporal change of variables x. Variables x are obtained as the summation of variables v by definition. We acquire the Pareto index, and the standard deviation σ of the growth rate distribution. It is confirmed that the two Pareto indices μ_v and μ_x are equal. It is found that σ exhibits scaling behavior with the scaling exponent ε. Because ε is negative, the standard deviation σ decreases as the size of x increases. This is consistent with the Second Non-Gibrat's Law. It is also worth noting that the scaling exponent ε depends on μ and α.

We have adopted only Eqs. (4) and (5) as the distribution of b. Other candidates should be considered, for example an exponential distribution. The effect of different choices should also be considered. It is also important to describe the relation between any one time step and actual time. Furthermore, analytic derivation of the statistical properties of x from the statistical properties of v is a significant issue. This study can serve as a clue to an analytic understanding of Non-Gibrat's Law. It might also provide a clue to knowledge of the management of credit risk.

Acknowledgments

The authors thank APFA7 & Tokyo Tech – Hitotsubashi Interdisciplinary Conference. Discussions during the Conference were useful to the completion of this work. The work was supported in part by a Grant-in-Aid for Scientific Research (C) (No. 20510147) from the Ministry of Education, Culture, Sports, Science and Technology, Japan. We also thank to P. M. Arathoon for his careful reading of the manuscript and his suggestions.

References

[1] Pareto V 1897 *Cours d'Economique Politique* (London: Macmillan)
[2] Axtell R L 2001 *Science* **7** 1818
[3] Gibrat R 1932 *Les inegalites economiques* (Paris: Sirey)
[4] Fujiwara Y, Souma W, Aoyama H, Kaizoji T and Aoki M 2003 *Physica* A **321** 59
Fujiwara Y, Guilmi C D, Aoyama H, Gallegati M and Souma W 2004 *Physica* A **335** 197
[5] Champernowne D G 1953 *Econ. Jour.* **63** 318
Simon H A and Bonini C P 1958 *Am. Econ. Rev.* **48** 607
Levy M and Solomon S 1996 *Int. J. Mod. Phys.* C **7** 595
[6] Stanley M H R, Amaral L A N, Buldyrev S V, Havlin S, Leschhorn H, Maass P, Salinger M A and Stanley H E 1996 *Nature* **379** 804
[7] Aoyama H 2004 *9th Ann. Workshop on Economic Heterogeneous Interacting Agents* (Kyoto)
[8] Aoyama H, Fujiwara Y and Souma W 2004 *The Physical Society of Japan 2004 Autumn Meeting* (Aomori)

[9] Ishikawa A 2006 *Physica* A **367** 425
[10] Ishikawa A 2007 *Physica* A **383** 79
[11] Takayasu H 2009 *APFA7& Tokyo Tech - Hitotsubashi Interdisciplinary Conf.* (Tokyo)
[12] Ishikawa A 2009 Power-Law and Log-Normal Distributions in Temporal Changes of Firm-Size Variables *Economics* **3** 2009-11
[13] Tomoyose M, Fujimoto S and Ishikawa A 2009 *Proc. of the YITP Workshop on Econophysics* (Kyoto) *Prog. Theor. Phys. Suppl.* **179** 114
[14] Takayasu H, Sato A and Takayasu M 1997 *Phys. Rev. Lett.* B **79** 966
[15] Ishikawa A, Fujimoto S and Tomoyose M 2009 submitted to *JEIC*
[16] Mizuno T, Takayasu M and Takayasu H 2004 *Physica* A **332** 403

Log-amplitude statistics of non-Gaussian fluctuations

Ken Kiyono

College of Engineering, Nihon University, 1 Naka-gawara, Tokusada, Tamura-machi, Koriyama City, Fukushima, 963-8642, Japan

E-mail: `kiyono@ge.ce.nihon-u.ac.jp`

Abstract. To quantify heavy-tailed distributions observed in financial time series, we propose a general method to characterize symmetric non-Gaussian distributions. In our approach, an observed time series is assumed to be described by the multiplication of Gaussian and amplitude random variables, where the amplitude variable describes fluctuations of the standard deviation. Based on this framework, it is shown that statistical properties of the log-amplitude fluctuations can be estimated using the logarithmic absolute moments of the observed time series.

1. Introduction

Recently, using statistical tools developed in nonequilibrium statistical physics, many studies have been conducted to establish universal properties in the financial market dynamics [1–3]. A remarkable property of the financial time serie is inhomogeneity of variance, which results in non-Gaussian probability density functions (PDFs). To characterize such non-Gaussian fluctuations, it has been demonstrated that non-Gaussian PDFs are often described effectively by a superposition of Gaussian distributions with fluctuating variances [4; 5].

In the study of the velocity difference between two points in fully developed turbulent flows, Castaing *et al.* [4] introduced the following PDF:

$$P(x) = \int_0^\infty \frac{1}{\sigma} P_L\left(\frac{x}{\sigma}\right) G(\ln\sigma)\, d(\ln\sigma), \tag{1}$$

where it is assumed that P_L is the Gaussian distribution and G is a distribution describing the fluctuation of the standard deviations. This PDF has been also applied to econophysics [1; 2], geophysics [6], and physiology [7]. However, in the previous studies, the estimation of the statistical properties of G depends on a priori knowledge of the functional form of $G(\ln\sigma)$ such as log-normality, which would limit the applicability of this model. Thus, it is required to develop a method for determining statistical properties of G from the observed data.

As pointed out by Jung and Swinney [8], equation (1) can be linked with Beck and Cohen's superstatistics [5] that has also been applied to a wide range of nonequilibrium systems (see, [5] and references therein). Hence, a general problem that is of great interest in experimental applications is how to objectively characterize the variance fluctuations described by G in Eq. (1). To solve this problem, we propose a novel method for estimating the variance and higher moments of G from the observed time series without assumptions on the shape of G.

2. Log-amplitude variance and higher moments

To explain our approach, let us assume that an observed stationary time series $\{x_i\}$ with zero mean is described by a multiplicative stochastic process,

$$X_i = W_i e^{Y_i}, \tag{2}$$

where W is a Gaussian random variable with zero mean, and Y is the other random variable independent of W. In this case, the PDF of X has the same functional form as Eq. (1), where $G(y)$ is the PDF of Y. Here, we refer to $\{Y_i\}$ as the log-amplitude fluctuation.

To characterize the log-amplitude fluctuation, we consider the variance and higher central moments of Y,

$$\mu_n = \langle (Y - \langle Y \rangle)^n \rangle , \tag{3}$$

where $\langle \cdot \rangle$ denotes the statistical average. By the calculation of the logarithmic absolute moments of X and the assumption of Eq. (2), we can obtain the following relations for μ_n:

$$\mu_2 = \left\langle (\ln|X| - \langle \ln|X| \rangle)^2 \right\rangle - \frac{\pi^2}{8}, \tag{4}$$

$$\mu_3 = \left\langle (\ln|X| - \langle \ln|X| \rangle)^3 \right\rangle + \frac{7}{4}\zeta(3), \tag{5}$$

$$\mu_4 = \left\langle (\ln|X| - \langle \ln|X| \rangle)^4 \right\rangle - \frac{3}{4}\pi^2 \mu_2 - \frac{7}{64}\pi^4, \tag{6}$$

where $\zeta(n)$ is the Riemann zeta function ($\zeta(3) = 1.2020569\cdots$). Note that the logarithmic absolute moments of X do not depend on the variance of X, because

$$\ln|\sigma_0 X| - \langle \ln|\sigma_0 X| \rangle = \ln|X| - \langle \ln|X| \rangle , \tag{7}$$

where σ_0 is an arbitrary constant. Furthermore, we can obtain the estimator of the higher-order moment, if required.

Our key idea is to calculate logarithmic absolute moments. If logarithmic absolute moments of X are finite, it is possible to define the log-amplitude moments of X. Even in the case where the PDF of X has power-law tails, $P(x) \sim |x|^{-\alpha}$, with $1 < \alpha$, the logarithmic absolute moments are finite, although the variance is undefined or infinite. Therefore, our approach can characterize a wide range of symmetric unimodal distributions with heavy tails. In other words, the log-amplitude variance μ_2 [Eq. (4)] is simply interpreted as the differnce in the second logarithmic-absolute moment between the observed non-Gaussian and Gaussian PDFs, because the second logarithmic-absolute moment of a Gaussian distribution equals $\pi^2/8$. Thus, μ_2 can be used as a measure of the deviation from a Gaussian distribution.

3. Numerical examples

To test our approach, we introduce illustrative examples of non-Gaussian stochastic processes and carry out numerical experiments. Here, we consider a stochastic process described by independent and identically distributed (IID) variables. The first example is a multiplicative log-normal process [9] based on experimental observations in the study of the turbulent velocity field [4], solar wind [6], foreign exchange rate [1], stock index [2], and human heartbeat [10; 11]. Neglecting the detailed structure of the intermittent dynamics, we mimic the PDFs observed in the log-normal processes. For comparison, we also introduce a multiplicative log-Poisson process. In the log-normal and log-Poisson processes, fluctuations of the standard deviations are assumed to obey log-normal and log-Poisson distributions, respectively.

In the log-normal process $\{X_i^{(\mathrm{LN})}\}$, the standardized random variable $X^{(\mathrm{LN})}$ with zero mean is described by

$$X^{(\mathrm{LN})} = e^{-\alpha^2} W e^{\alpha Y}, \tag{8}$$

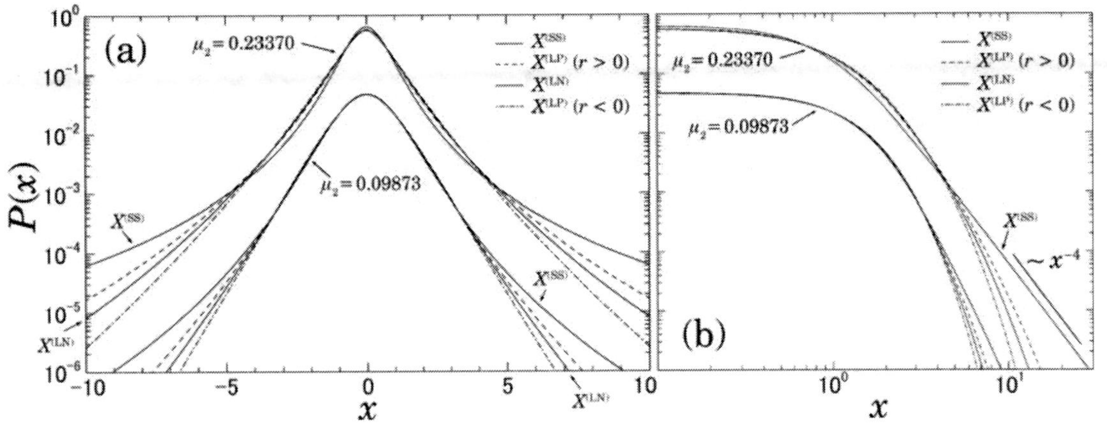

Figure 1. (a) Standardized PDFs of log-normal ($X^{(\mathrm{LN})}$), log-Poisson ($X^{(\mathrm{LP})}$), and superstatistical ($X^{(\mathrm{SS})}$) IID processes with the same log-amplitude variance μ_2, where $X^{(\mathrm{LN})}$, $X^{(\mathrm{LP})}$, and $X^{(\mathrm{SS})}$ are defined by Eqs. (8), (9), and (12), respectively. Solid lines: $X^{(\mathrm{SS})}$ with $k = 3$ (top) and $k = 6$ (bottom); dashed lines: $X^{(\mathrm{LP})}$ with $\lambda = 20$ and $r = \sqrt{\mu_2/\lambda}$; solid lines: $X^{(\mathrm{LN})}$ with $\alpha = \sqrt{\mu_2}$; dot-dashed lines: $X^{(\mathrm{LP})}$ with $\lambda = 20$ and $r = -\sqrt{\mu_2/\lambda}$. The PDFs are shifted in vertical directions for convenience of presentation. (b) Double-logarithmic plot of (a).

where both W and Y are independent standard Gaussian random variables with zero mean and unit variance. In this process, the non-Gaussian nature is determined by the parameter α, which is the same as a non-Gaussian parameter λ defined in Ref. [9]. In this case, its log-amplitude moments are $\mu_2 = \alpha^2$, $\mu_3 = 0$ and $\mu_4 = 3\alpha^4$.

In the log-Poisson process $\{X_i^{(\mathrm{LP})}\}$, the standardized random variable $X^{(\mathrm{LP})}$ is described by

$$X^{(\mathrm{LP})} = \exp\left(-\lambda\left(\exp(2r) - 1\right)/2\right) W e^{rP}, \tag{9}$$

where W are independent standard Gaussian random variables, P are independent Poisson random variables with mean λ and variance λ, r is a real valued parameter. In this case, its log-amplitude moments are $\mu_2 = r^2\lambda$, $\mu_3 = r^3\lambda$, and $\mu_4 = r^4\lambda(1 + 3\lambda)$.

The next example is a stochastic process $\{X_i^{(\mathrm{SS})}\}$ based on so-called superstatistics [5]. Superstatistics considers an inhomogeneous driven nonequilibrium system that consists of many subsystems with different values of some intensive parameter β. Each subsystem is assumed to reach local equilibrium very quickly. In this case, if the local equilibrium distribution is Gaussian, we obtain

$$P(x) = \int_0^\infty \sqrt{\beta}\, P_L\left(\sqrt{\beta} x\right) f(\beta)\, d\beta, \tag{10}$$

where $P_L(x)$ is the standard normal distribution and $f(\beta)$ is the distribution of β. If $f(\beta)$ is a log-normal distribution, the PDF of superstatistics has the same form as the above log-normal process [8].

Here, we choose the χ^2 distribution with degree k,

$$f(\beta) = \frac{1}{\Gamma(k/2)} \left(\frac{k}{2\beta_0}\right)^{k/2} \beta^{\frac{k}{2}-1} e^{-\frac{k\beta}{2\beta_0}}, \tag{11}$$

which is one of the universality classes proposed in superstatistics [5]. In this case, equation (10) results in the Student's t-distribution, which exhibits power-law tails, $P(x) \sim |x|^{-(k+1)}$ for

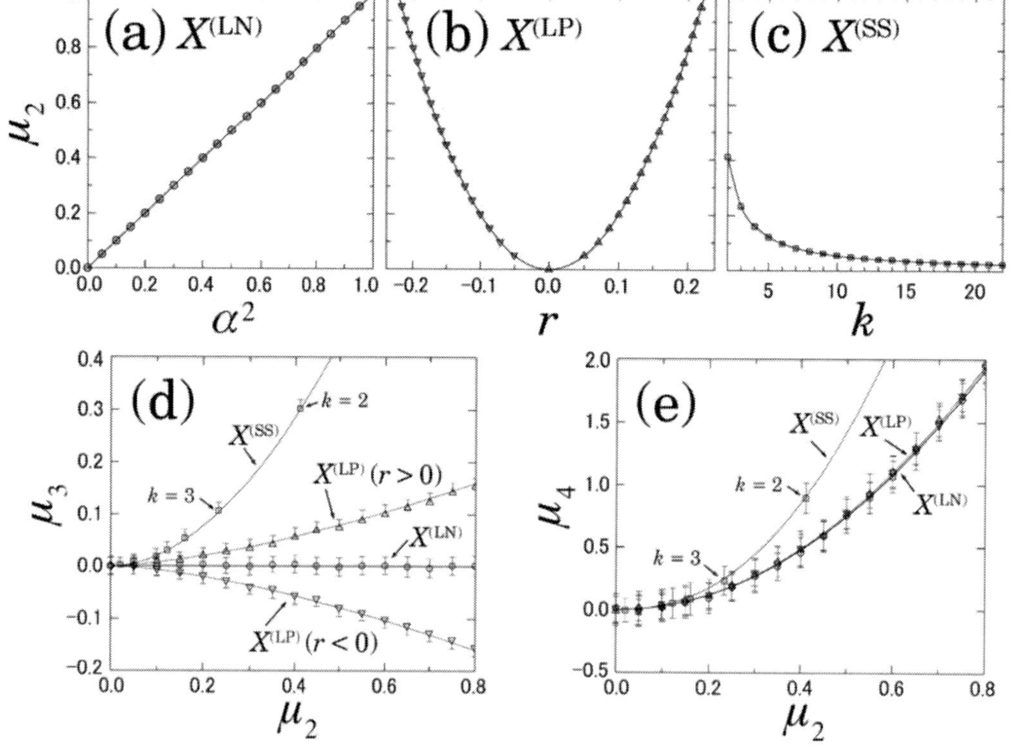

Figure 2. Estimation of μ_n for log-normal ($X^{(\mathrm{LN})}$, circles), log-Poisson ($X^{(\mathrm{LP})}$, triangles), and superstatistical ($X^{(\mathrm{SS})}$, squares) IID processes, where $\lambda = 20$ for $X^{(\mathrm{LP})}$. The sample means of μ_n were estimated from 100 samples of length $N = 10^6$. The error bars indicate the sample standard deviation. The solid lines indicate the theoretical predictions.

large $|x|$. In this superstatistical process, the random variable $X^{(\mathrm{SS})}$ is described as

$$X^{(\mathrm{SS})} = W\sqrt{\frac{k}{\beta_0 Q}} \tag{12}$$

where W are independent standard Gaussian random variables, Q are independent χ^2 random variables with k degrees of freedom. When $k > 2$, $X^{(\mathrm{SS})}$ can be standardized by $\beta_0 = k/(k-2)$. In Castaing's description [Eq. (1)], the corresponding $G(y)$ is defined by

$$G(y) = \frac{2}{\Gamma(k/2)} \exp\left\{-k\left(y - \frac{1}{2}\ln\frac{k}{2\beta_0}\right) - e^{-2\left(y - \frac{1}{2}\ln\frac{k}{2\beta_0}\right)}\right\}, \tag{13}$$

which is the Gumbel distribution when $k = 2$. In this case, its log-amplitude moments are $\mu_2 = \psi^{(1)}(k/2)/4$, $\mu_3 = -\psi^{(2)}(k/2)/8$, and $\mu_4 = (3\,\psi^{(1)}(k/2)^2 + \psi^{(3)}(k/2))/16$, where $\psi^{(n)}(x)$ is nth derivative of the Euler's psi function $\psi(x)$.

It is important to note that the log-amplitude moments μ_n can be defined for $k > 0$. When $k = 1$, the PDF of $X^{(\mathrm{SS})}$ reduces to a Cauchy distribution,

$$P(x) = \frac{\gamma}{\pi\,(x^2 + \gamma^2)}, \tag{14}$$

where the scale parameter γ is chosen as $\gamma = 1/\sqrt{\beta_0}$. Because this distribution has power-law tails, $P(x) \sim |x|^{-2}$, for large $|x|$, its second and higer moments are infinite. On the other hand, all of the log-amplitude moments μ_n are finite, which demonstrates that the log-amplitude statistics is applicable to a variety of heavy tailed distributions.

As shown in Fig. 1, the center parts of the PDFs have similar shapes, if the values of μ_2 are equal. To test our approach, we numerically generate data sets using the above models, and then estimate the value of μ_n. To estimate the logarithmic absolute moments, $\langle(\ln|X| - \langle\ln|X|\rangle)^n\rangle$, in Eqs. (4)-(6), here we use the following estimators,

$$M_n = \frac{1}{N}\sum_{i=1}^{N}(\ln|X_i| - M_1)^n \quad (n = 2, 3, 4), \tag{15}$$

where

$$M_1 = \frac{1}{N}\sum_{i=1}^{N}\ln|X_i|. \tag{16}$$

As shown in Fig. 2, the theoretical values are estimated well from the observed time series. In particular, in the plot of μ_3 vs μ_2 [Fig. 2 (d)], we find significant differences between the models, although in the plot of μ_4 vs μ_2 [Fig. 2 (e)] we find no differences between the log-normal and log-Poisson processes.

4. Conclusion

We proposed log-amplitude statistics to characterize non-Gaussian time series. Both turbulence statistics by Castaing *el al.* [4] and superstatistics by Beck and Cohen [5] have been very successful in describing non-Gaussian fluctuations. Including such examples, our method can be used to characterize non-Gaussian fluctuations.

As discussed in the previous section, a crucial advantage in our approach is that a priori knowledge of the variance fluctuations is not assumed. Hence, its increasingly widespread application in the analysis of financial time series is foreseen.

References

[1] Ghashghaie S, Breymann W, Peinke J, Talkner P and Dodge Y 1996 *Nature* **381** 767–770

[2] Kiyono K, Struzik Z R and Yamamoto Y 2006 *Phys. Rev. Lett.* **96** 068701

[3] Muzy J F, Delour J and Bacry E 2000 *Eur. Phys. J. B* **17** 537–548

[4] Castaing B, Gagne Y and Hopfinger E J 1990 *Physica D* **46** 177–200

[5] Beck C and Cohen E G D 2003 *Physica A* **322** 267–275

[6] Sorriso-Valvo L, Carbone V, Veltri P, Consolini G and Bruno R 1999 *Geophys. Res. Lett.* **26** 1801–1804

[7] Struzik Z R, Kiyono K, Hayano J, Watanabe E and Yamamoto Y 2008 *Europhys. Lett.* **82** 28005

[8] Jung S and Swinney H L 2005 *Phys. Rev. E* **72** 026304

[9] Kiyono K, Struzik Z R and Yamamoto Y 2007 *Phys. Rev. E* **76** 041113

[10] Kiyono K, Struzik Z R, Aoyagi N, Sakata S, Hayano J and Yamamoto Y 2004 *Phys. Rev. Lett.* **93** 178103

[11] Kiyono K, Struzik Z R, Aoyagi N, Togo F and Yamamoto Y 2005 *Phys. Rev. Lett.* **95** 058101

Bayesian inference with an adaptive proposal density for GARCH models

Tetsuya Takaishi

Hiroshima University of Economics, Hiroshima 731-0192, JAPAN

E-mail: `takaishi@hiroshima-u.ac.jp`

Abstract. We perform the Bayesian inference of a GARCH model by the Metropolis-Hastings algorithm with an adaptive proposal density. The adaptive proposal density is assumed to be the Student's t-distribution and the distribution parameters are evaluated by using the data sampled during the simulation. We apply the method for the QGARCH model which is one of asymmetric GARCH models and make empirical studies for Nikkei 225, DAX and Hang indexes. We find that autocorrelation times from our method are very small, thus the method is very efficient for generating uncorrelated Monte Carlo data. The results from the QGARCH model show that all the three indexes show the leverage effect, i.e. the volatility is high after negative observations.

1. Introduction

In empirical finance volatility of asset returns is an important value to measure risk. In order to forecast future volatility it is desirable to use appropriate models which have the properties of volatility of asset returns. Many empirical studies suggest that the distribution of asset returns is leptokurtic. Furthermore the volatility is not constant, but changes over time. There are periods when volatility is very high or very low. This property of the volatility is called *volatility clustering.*

The Autoregressive Conditional Heteroscedasticity (ARCH) model[1] and its generalization, the Generalized ARCH (GARCH) model[2] are designed to capture the property of the volatility clustering. Moreover the distributions of returns from those models show fat-tailed distributions and they are suggested to be q-Gaussian distributions[3, 4]. There are many extensions of the GARCH model to include additional properties of the volatility. An example of the properties of the volatility is that the volatility response is high after negative news (returns), which is known as the leverage effect, first observed by Black[5]. In order to cope with this fact, some models[6, 7, 8, 9, 10, 11] which introduce asymmetry into the volatility response function are proposed. In this study among them we focus on the Quadratic GARCH(QGARCH) model[10, 11] which adds an additional term proportional to a return to the volatility response function.

To utilize the GARCH models we need to infer GARCH parameters from financial time series data. In general the Maximum Likelihood (ML) estimation is favored to the inference of GARCH models. Although implementation of the ML method is straightforward, there exist practical difficulties in estimating GARCH parameters by the ML technique. The model parameters must be positive to ensure a positive volatility and the stationarity condition for volatility is also required. The ML method with such requirements is performed via a constrained optimization technique which can be cumbersome. Forethermore the output of the ML method is often sensitive to starting values.

Another estimation technique is the Bayesian inference which does not have the difficulties seen in the ML method. Usually the Bayesian inference is performed by Markov Chain Monte Carlo (MCMC) methods which have been common in the recent computer development. Various MCMC methods for the Bayesian inference of the GARCH models have been proposed[12, 13, 14, 15, 16, 17, 18, 19]. In a survey on the MCMC methods of the GARCH models[17] it is shown that Acceptance-Rejection/ Metropolis-Hastings (AR/MH) algorithm with a multivariate Student's t-distribution works better than the other algorithms. The multivariate Student's t-distribution is used as a proposal density of the MH algorithm and the parameters to specify the Student's t-distribution are determined by the Maximum Likelihood (ML) technique. Recently an alternative method to estimate those parameters without relying on the ML technique was proposed[20, 21, 22]. The method is called "adaptive construction scheme", where the parameters of the multivariate Student's t-distribution are determined by using the pre-sampled data by an MCMC method. And the parameters are updated adaptively during the MCMC simulation.

The adaptive construction scheme was tested for GARCH and QGARCH models[20, 21, 22] and it is shown that the adaptive construction scheme can significantly reduce the correlation between sampled data. In this paper first we describe the adaptive construction scheme for the GARCH models. Then we make empirical studies with the QGARCH model for three major stock indexes, Nikkei 225, DAX and Hang Seng.

2. GARCH Model

Let x_t be an asset return observed at time t. We transform x_t to y_t as

$$y_t = x_t - \bar{x}, \tag{1}$$

where $\bar{x}$ is the average over N observations, i.e. $\bar{x} = \frac{1}{N}\sum_{i=1}^{N} x_i$. In the GARCH model y_t is assumed to be decomposed as

$$y_t = \sigma_t \epsilon_t, \tag{2}$$

where ϵ_t is an independent and identically distributed (i.i.d.) random variable with zero mean and unit variance. The distribution of ϵ_t is usually assumed to be a normal (Gaussian) or a leptokurtic one. In this study we assume that the distribution is a normal one, i.e. $\sim N(0,1)$. In the original GARCH model the volatility σ_t is assumed to change over time as

$$\sigma_t^2 = \omega + \sum_{i=1}^{q} \alpha_i y_{t-i}^2 + \sum_{i=1}^{p} \beta_i \sigma_{t-i}^2, \tag{3}$$

and more specifically this model is stated as the GARCH(p,q) model. Since in empirical studies with the AIC analysis small numbers are often chosen for p and q we focus on the GARCH(1,1) model in this study and write it as the GARCH model. Now eq.(3) is written as

$$\sigma_t^2 = \omega + \alpha y_{t-1}^2 + \beta \sigma_{t-1}^2. \tag{4}$$

The volatility response function of eq.(4) is symmetric under positive or negative observations y_t. However some asset returns are known to show the leverage effect, i.e. the volatility is higher after negative observations than after positive ones. Several extended GARCH models have been proposed to introduce asymmetry into the volatility response function. In this study we use the QGARCH model[10, 11] given by

$$\sigma_t^2 = \omega + \gamma y_{t-1} + \alpha y_{t-1}^2 + \beta \sigma_{t-1}^2, \tag{5}$$

where the additional term, γy_{t-1} introduces the asymmetry into the model. The QGARCH model includes 4 parameters $(\omega, \alpha, \beta, \gamma)$ which have to be determined from financial data.

The positivity of eq.(5) is always ensured if the parameter space is restricted to $\alpha > 0$, $\beta > 0$ and $\omega - \gamma^2/(4\alpha) > 0$. Later we see that our final outputs of the parameters (Table 1) satisfies these conditions. There may be a possibility that in the MCMC simulations these conditions are weakly violated depending on the current values of the parameters. However empirically we observed no negative σ_t^2 during MCMC simulations for the financial data we used.

3. Maximum Likelihood Estimation

The likelihood function of the GARCH models is written as

$$L(y|\theta) = \Pi_{i=1}^{n} \frac{1}{\sqrt{2\pi\sigma_t^2}} \exp\left(-\frac{y_t^2}{2\sigma_t^2}\right), \tag{6}$$

where y stands for the time series data of n observations, $y = (y_1, \ldots .y_n)$ and θ stands for GARCH parameters, e.g. for the QGARCH model $\theta = (\omega, \alpha, \beta, \gamma)$.

In the ML estimation the parameters are determined by maximizing the log likelihood $\ln L(y|\theta)$,

$$\ln L(y|\theta) = -\frac{1}{2}\sum_i^n \ln(2\pi\sigma_t^2) - \sum_i^n \frac{y_t^2}{2\sigma_t^2}. \tag{7}$$

4. Bayesian Inference

From the Bayes' theorem we obtain the posterior density $\pi(\theta|y)$ which is a probability distribution of parameters θ as

$$\pi(\theta|y) \propto L(y|\theta)\pi(\theta), \tag{8}$$

where $\pi(\theta)$ is the prior density for θ. Since we do not know the functional form of $\pi(\theta)$ we make an assumption for it. Here we assume that $\pi(\theta)$ is constant. Using $\pi(\theta|y)$, θ is estimated by

$$\langle\theta\rangle = \frac{1}{Z}\int \theta\pi(\theta|y)d\theta, \tag{9}$$

where

$$Z = \int \pi(\theta|y)d\theta. \tag{10}$$

In general eq.(9) can not be performed analytically. Instead of performing the integral of eq.(9) we evaluate eq.(9) by the MCMC technique described in the next section. In the MCMC evaluation the normalization constant Z will be irrelevant.

5. Markov Chain Monte Carlo

The Metropolis algorithm is an MCMC technique first introduced by Metropolis *et al.*[23], which is designed to estimate an integral such as eq.(9) numerically. The Metropolis-Hastings (MH) algorithm[24] is a generalization of the original Metropolis algorithm. Let us consider to evaluate eq.(9) by the MH algorithm. In general the functional form of $\pi(\theta|y)$ may be too complicated to generate random variables according to $\pi(\theta|y)$. In the MH algorithm we draw random variables from a proposal density which is simple enough to generate random variables. The basic procedure of the MH algorithm is given as follows.

(1) First we set an initial value θ_0 and $i = 1$.

(2) Then we generate a new value θ_i from a certain proposal density $g(\theta_i|\theta_{i-1})$.

(3) We accept the candidate θ_i with a probability of $P_{MH}(\theta_{i-1}, \theta_i)$ where

$$P_{MH}(\theta_{i-1}, \theta_i) = \min\left[1, \frac{P(\theta_i)}{P(\theta_{i-1})} \frac{g(\theta_{i-1}|\theta_i)}{g(\theta_i|\theta_{i-1})}\right]. \tag{11}$$

When θ_i is rejected we keep θ_{i-1}, i.e. $\theta_i = \theta_{i-1}$.

(4) Go back to (2) with an increment of $i = i + 1$.

When the proposal density does not depend on the previous value, i.e. $g(\theta_i|\theta_{i-1}) = g(\theta_i)$ we obtain

$$P_{MH}(\theta_{i-1}, \theta_i) = \min\left[1, \frac{P(\theta_i)}{P(\theta_{i-1})} \frac{g(\theta_{i-1})}{g(\theta_i)}\right]. \tag{12}$$

For a symmetric proposal density $g(\theta_i|\theta_{i-1}) = g(\theta_{i-1}|\theta_i)$, eq.(11) reduces to the Metropolis algorithm and the Metropolis accept probability is given by

$$P_{Metro}(\theta_{i-1}, \theta_i) = \min\left[1, \frac{P(\theta_i)}{P(\theta_{i-1})}\right]. \tag{13}$$

Eq.(9) is evaluated as an average over the data sampled by the MCMC algorithm.

$$\langle\theta\rangle = \lim_{N\to\infty} \frac{1}{k} \sum_{i=1}^{N} \theta^{(i)}, \tag{14}$$

where N is the number of the sampled data. For N independent data the statistical error is proportional to $\frac{1}{\sqrt{N}}$. However the data sampled by MCMC methods are not independent. For N correlated data, the statistical error is proportional to $\sqrt{\frac{2\tau}{N}}$ where τ is the autocorrelation time between the sampled data. Thus in order to have a small statistical error without increasing the number of sampled data, it is important to take an MCMC method which generates uncorrelated data.

6. Adaptive Construction Scheme

The efficiency of the MH algorithm depends on how we choose the proposal density. By choosing an adequate proposal density for the MH algorithm one can reduce the correlation between the sampled data. The posterior density of GARCH parameters often resembles to a Gaussian-like shape. Thus one may choose a density similar to a Gaussian distribution as the proposal density. Such attempts have been done by Mitsui, Watanabe[16] and Asai[17]. They used a multivariate Student's t-distribution in order to cover the tails of the posterior density and determined the parameters to specify the distribution by using the ML technique. Here we also use a multivariate Student's t-distribution but determine the parameters through MCMC simulations.

The (p-dimensional) multivariate Student's t-distribution is given by

$$\begin{aligned} g(\theta) &= \frac{\Gamma((\nu+p)/2)/\Gamma(\nu/2)}{\det \Sigma^{1/2}(\nu\pi)^{p/2}} \\ &\times \left[1 + \frac{(\theta - M)^t \Sigma^{-1} (\theta - M)}{\nu}\right]^{-(\nu+p)/2}, \end{aligned} \tag{15}$$

where θ and M are column vectors,

$$\theta = \begin{bmatrix} \theta_1 \\ \theta_2 \\ \vdots \\ \theta_p \end{bmatrix}, M = \begin{bmatrix} M_1 \\ M_2 \\ \vdots \\ M_p \end{bmatrix}, \tag{16}$$

and $M_i = E(\theta_i)$. Σ is the covariance matrix defined as

$$\frac{\nu\Sigma}{\nu - 2} = E[(\theta - M)(\theta - M)^t]. \tag{17}$$

ν is a parameter to tune the shape of Student's t-distribution. When $\nu \to \infty$ the Student's t-distribution goes to a Gaussian distribution. At small ν Student's t-distribution has a fat-tail. Since eq.(15) is independent of the previous value of θ, eq.(12) is used in the MH algorithm.

To use eq.(15) as a proposal density we have to know the values of M and Σ. We determine these unknown parameters M and Σ as follows. First we make a short run by an MCMC method and sample some data. Then we estimate M and Σ from those data. Second substituting the estimated M and Σ to eq.(15) we perform an MH simulation with the proposal density. After accumulating more data, we recalculate M and Σ, and update M and Σ of eq.(15). By doing this, we adaptively change the shape of eq.(15) to fit the posterior density more accurately. We call eq.(15) constructed in this way "adaptive proposal density".

The random number generation for the multivariate Student's t-distribution can be done easily as follows. First we decompose the symmetric covariance matrix Σ by the Cholesky decomposition as $\Sigma = LL^t$. Then substituting this result to eq.(15) we obtain

$$g(X) \sim \left[1 + \frac{X^t X}{\nu}\right]^{-(\nu+p)/2}, \tag{18}$$

where $X = L^{-1}(\theta - M)$. The random numbers X are given by $X = Y\sqrt{\frac{\nu}{w}}$, where Y follows $N(0, I)$ and w is taken from the chi-square distribution ν degrees of freedom χ^2_ν. Finally we obtain the random number θ by $\theta = LX + M$.

7. Empirical Studies

In this section we make empirical studies based on daily data of Nikkei 225, DAX and Hang Seng indexes. The sampling period is 4(2)(3) January 1995 to 30 December 2005 for the Nikkei 225 (DAX)(Hang Seng) index. The index prices p_i are transformed to returns as

$$r_i = 100 \ln(p_i/p_{i-1} - \bar{s}), \tag{19}$$

where $\bar{s}$ is the average value of $\ln(p_i/p_{i-1})$. Fig.1 shows the time series of returns of the Nikkei 225 index calculated by eq.(19) as an example.

The adaptive construction scheme is implemented as follows. First we make a short run by an MCMC method. Any MCMC method can be used. Here we use the Metropolis algorithm. We discard the first 5000 data as burn-in process[1]. Then we accumulate 1000 data to estimate M and Σ. The estimated M and Σ are substituted to $g(\theta)$ of eq.(15). The shape parameter ν is set to 10. We re-start a run by the MH algorithm with the proposal density $g(\theta)$. Every 1000

[1] This process is also called "thermalization".

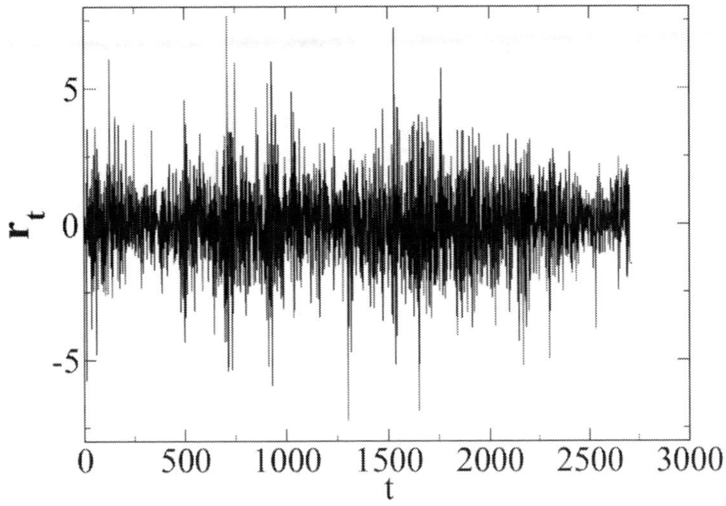

Figure 1. Time series of returns of the Nikkei 225 index.

update we re-calculate M and Σ using all accumulated data and update $g(\theta)$ for the next run. We accumulate 100000 data for analysis. The results analyed are summarized in Table 1.

We examine the efficiency of the algorithm by measuring correlations between sampled data. Fig.2 shows the autocorrelation function (ACF) of the sampled α with the Nikkei 225 index data. The ACF is defined as

$$ACF(t) = \frac{\frac{1}{N}\sum_{j=1}^{N}(x(j) - \langle x \rangle)(x(j+t) - \langle x \rangle)}{\sigma_x^2}, \tag{20}$$

where $\langle x \rangle$ and σ_x^2 are the average value and the variance of certain successive data x respectively. The ACF quickly decreases with t, which indicates that the correlation between the sampled data is very small. We also find the similar behavior for the other parameters.

In order to analyze the autocorrelation quantitatively we measure the integrated autocorrelation time τ_{int} given by

$$\tau_{int} = \frac{1}{2} + \sum_{i=1}^{\infty} ACF(i). \tag{21}$$

"$2\tau_{int}$" is also called inefficiency factor(see, e.g., [25]). If no correlation exists in the data $2\tau_{int}$ takes one. The results of $2\tau_{int}$ are summarized in Table 1. We find that all the $2\tau_{int}$ are very small, less than 2, which indicates that the correlation between the data sampled by this method is small.

Fig.3 and 4 show the convergence property of the covariance matrix. Here V is a 4×4 matrix defined by $V = E[(\theta - M)(\theta - M)^t]$. The elements V_{11} and V_{12} quickly converge to certain values. We also see the similar behavior for the other elements.

Fig.5 shows the acceptance at the MH algorithm with the adaptive proposal density of eq.(15). Each acceptance is calculated every 1000 updates and the calculation of the acceptance is based on the latest 1000 data. At the first stage of the simulation the acceptance is low. This is because M and Σ have not yet been calculated accurately at this stage as seen in figs. 3-4. The acceptance increases quickly as the simulation is proceeded and reaches plateaus of about 80%.

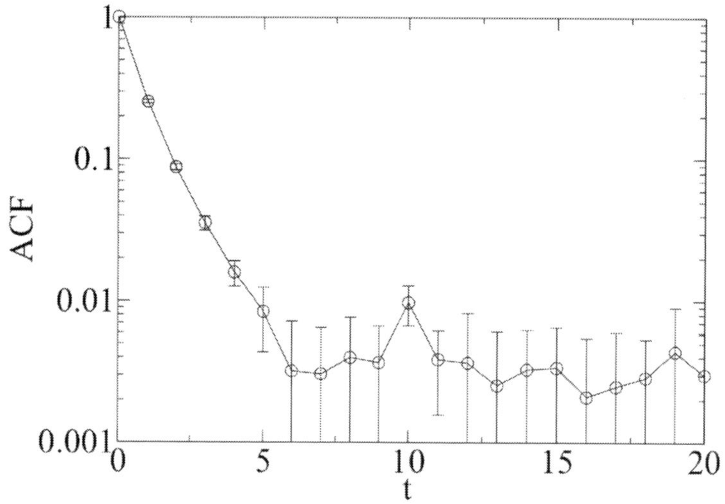

Figure 2. Autocorrelation functions of α sampled with the Nikkei 225 index data.

Table 1. Results of QGARCH parameters. SD and SE stand for standard deviation and statistical error respectively. Theoretically there is a relation of $SE \approx \sqrt{\frac{2\tau_{int}}{N}}SD$. Here statistical errors are estimated by the jackknife method.

	α	β	ω	γ
Nikkei 225	0.07872	0.89390	0.06219	-0.12403
SD	0.011	0.013	0.013	0.021
SE	0.00003	0.00004	0.00005	0.00007
$2\tau_{int}$	2.0 ± 0.1	2.0 ± 0.1	2.0 ± 0.2	1.8 ± 0.1
DAX	0.09198	0.89564	0.03004	-0.08483
SD	0.011	0.011	0.0064	0.015
SE	0.00004	0.00005	0.00004	0.00006
$2\tau_{int}$	1.77 ± 0.06	1.78 ± 0.05	1.80 ± 0.07	1.61 ± 0.06
Hang Seng	0.07638	0.91168	0.03202	-0.08678
SD	0.009	0.0098	0.007	0.007
SE	0.00005	0.00005	0.00003	0.00007
$2\tau_{int}$	1.80 ± 0.06	1.78 ± 0.06	1.79 ± 0.06	1.75 ± 0.06

The QGARCH model is capable of capturing leverage effects. In order to see the impact of leverage effects Pagan and Schwert[26], and Engle and Ng[27] proposed the use of the so-called news impact curve. The news impact curve is the functional relationship between conditional variance at time t and the shock at time $t-1$, y_{t-1}.

The news impact curve of the QGARCH model is given by

$$\sigma_t^2 = \omega + \gamma y_{t-1} + \alpha y_{t-1}^2 + \beta\sigma^2, \tag{22}$$

where σ^2 is the unconditional variance given by

$$\sigma^2 = \frac{\omega}{1-\alpha-\beta}. \tag{23}$$

Fig.6 shows the impact curves of Nikkei 225, DAX and Hang Seng indexes. Since the values of γ for three indexes are all negative the three indexes exhibit the leverage effect. As seen in

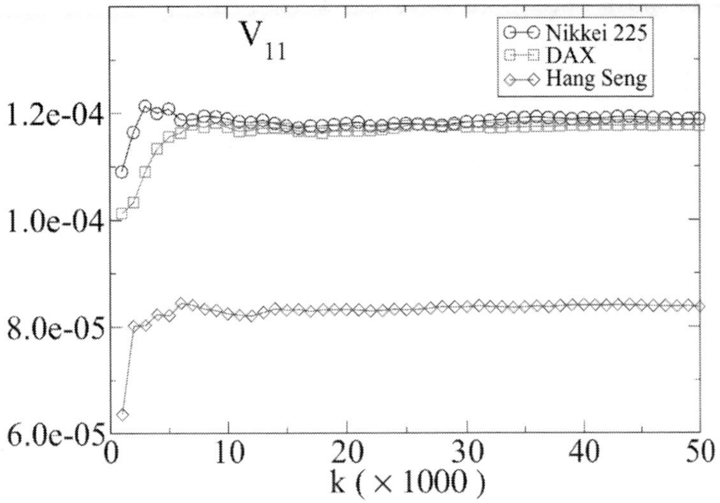

Figure 3. The diagonal element V_{11} as a function of Monte Carlo time.

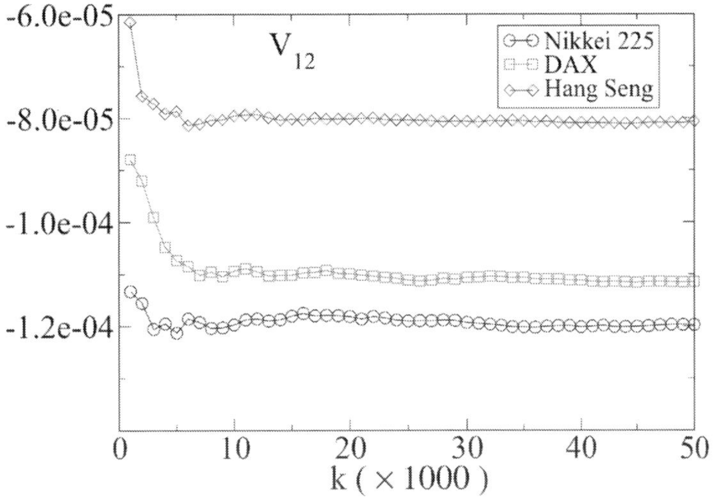

Figure 4. The off-diagonal element V_{12} as a function of Monte Carlo time.

fig.6 the news impact curve is higher for negative y_{t-1} than for positive one. Since all the three news impact curves are very similar each other the three indexes turn out to have the similar response to positive or negative news.

8. Conclusions

We have performed the Bayesian inference of the QGARCH model by the MCMC algorithm. The MCMC algorithm was implemented by the MH method with the adaptive proposal density. The adaptive proposal density is assumed to be the Student's t-distribution and the distribution parameters are determined by the data sampled by the MCMC simulation. The distribution

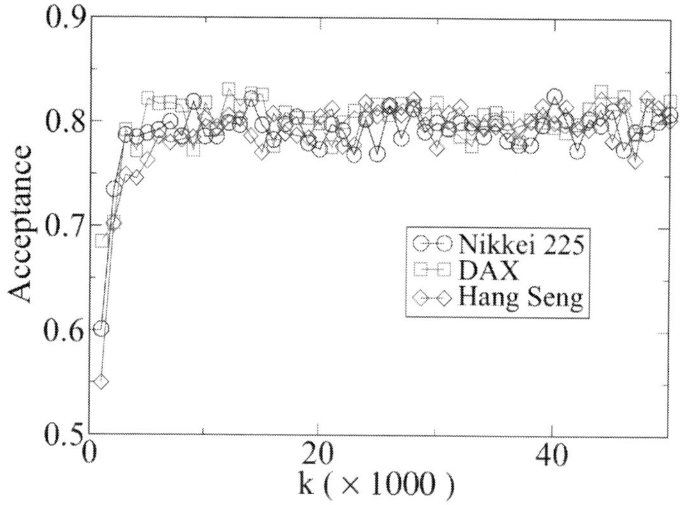

Figure 5. Acceptance at the MH algorithm with the adaptive proposal density.

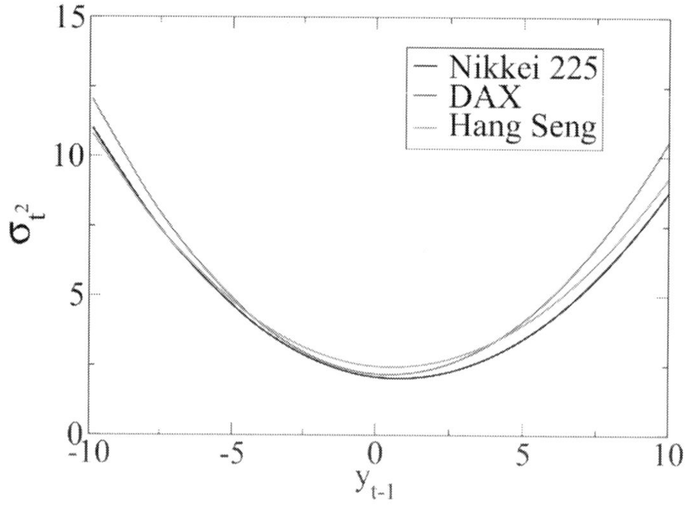

Figure 6. News impact curves of the Nikkei 225, DAX and Hang Seng indexes.

parameters are updated during the MCMC simulation adaptively to match the posterior density of the model parameters.

We have applied our method for Nikkei 225, DAX and Hang Seng indexes. We find that the autocorrelation times between the sampled data are very small, typically $2\tau_{int}$ is less than 2. Thus our method is very efficient for generating uncorrelated Monte Carlo data.

The QGARCH model is designed to capture the leverage effect. We find that all the three indexes show the leverage effect, i.e. the volatility is higher after negative observations than after positive ones.

The acceptance of the MH algorithm quickly reaches a plateau of about 80% already at the beginning of the simulation. This means that the distribution parameters are evaluated accurately already at the beginning of the simulation, as shown in figs.3-4. This observation suggests that in practice one may stop the update of the parameters at some stage of the simulation.

Acknowledgments
The numerical calculations were carried out on SGI Altix3700 at the Institute of Statistical Mathematics and on NEC SX8 at the Yukawa Institute for Theoretical Physics in Kyoto University. This study was carried out under the ISM Cooperative Use Registration (2009-ISM-CUR-0005).

References
[1] Engle R F 1982 Autoregressive Conditional Heteroskedasticity with Estimates of the Variance of the United Kingdom inflation *Econometrica* **60** 987
[2] Bollerslev T 1986 Generalized Autoregressive Conditional Heteroskedasticity *Journal of Econometrics* **31**, 307–327 (1986)
[3] Duarte Queiros S M and Tsallis C 2005 Bridging the ARCH model for finance and nonextensive entropy *Europhys. Lett.* **69** 893
[4] Duarte Queiros S M and Tsallis C 2005 On the connection between financial processes with stochastic volatility and nonextensive statistical mechanics *Eur. Phys. J. B* **48** 139
[5] Black F 1976 Studies of Stock Market Volatility Changes *1976 Proceedings of the American Statistical Association, Business and Economic Statistics Section* 177
[6] Nelson D B 1991 Conditional Heteroskedasticity in Asset Returns: A New Approach *Econometrica* **59**, 347
[7] Glosten L R, Jaganathan R and Runkle D E 1993 On the Relation Between the Expected Value and the Volatility of the Nominal Excess on Stocks *Journal of Finance* **48** 1779
[8] Ding Z, Granger C W J and Engle R F 1993 A Long Memory Property of Stock Market Returns and a New Model *Journal of Empirical Finance* **1**, 83
[9] Zakoïan M 1994 Threshold Heteroscedastic Models *Journal of Economic Dynamics and Control* **18** 931
[10] Engle R F and Ng V 1993 Measuring and testing the impact of news on volatility *Journal of Finance* **48** 1749
[11] Sentana E 1995 Quadratic ARCH models *Review of Economic Studies* **62** 639
[12] Bauwens L and Lubrano M 1998 Bayesian inference on GARCH models using the Gibbs sampler *Econometrics Journal* **1** c23
[13] Kim S Shephard N and Chib S 1998 Stochastic volatility: Likelihood inference and comparison with ARCH models *Review of Economic Studies* **65** 361
[14] Nakatsuma T 2000 Bayesian analysis of ARMA-GARCH models: Markov chain sampling approach *Journal of Econometrics* **95** 57
[15] Vrontos I D, Dellaportas P and Politis D N 2000 Full Bayesian inference for GARCH and EGARCH models *Journal of Business and Economic Statistics* **18** 187
[16] Mitsui H and Watanabe T 2003 Bayesian analysis of GARCH option pricing models *J. Japan Statist. Soc. (Japanese Issue)* **33** 307
[17] Asai M 2006 Comparison of MCMC Methods for Estimating GARCH Models *J. Japan Statist. Soc.* **36** 199
[18] Takaishi T 2006 Bayesian Estimation of GARCH model by Hybrid Monte Carlo Proceedings of the 9th Joint Conference on Information Sciences 2006 CIEF-214
doi:10.2991/jcis.2006.159
[19] Ardia D 2008 Financial Risk Management with Bayesian Estimation of GARCH models (Springer)
[20] Takaishi T 2009 An Adaptive Markov Chain Monte Carlo Method for GARCH Model *Lecture Notes of the Institute for Computer Sciences, Social Informatics and Telecommunications Engineering, Complex Sciences* Vol. **5** 1424
doi:10.1007/978-3-642-02469-6_22
[21] Takaishi T 2009 Bayesian Estimation of GARCH Model with an Adaptive Proposal Density *New Advances in Intelligent Decision Technologies, Studies in Computational Intelligence* Vol. **199** 635
doi:10.1007/978-3-642-00909-9_61
[22] Takaishi T 2009 Bayesian Inference on QGARCH Model Using the Adaptive Construction Scheme *Proceedings of 8th IEEE/ACIS International Conference on Computer and Information Science* 525
doi:10.1109/ICIS.2009.173

[23] Metropolis N *et al.* 1953 Equations of State Calculations by Fast Computing Machines *J. of Chem. Phys.* **21** 1087

[24] Hastings W K 1970 Monte Carlo Sampling Methods Using Markov Chains and Their Applications *Biometrika* **57** 97

[25] Chib S 2001 Markov chain Monte Carlo methods: computation and inference *In: Handbook of Econometrics* Vol. **5** Amsterdam, North Holland, 3569–3649.

[26] Pagan A and Schwert G W 1990 Alternative Models for Conditional Volatility *Journal of Econometrics* **45** 267

[27] Engle R F and Ng V 1993 Measuring and Testing the Impact of News on Volatility *Journal of Finance* **48** 1749

COMMUNITY STRUCTURE IN A LARGE-SCALE TRANSACTION NETWORK AND VISUALIZATION

T. Iino[a,*)], K. Kamehama[a], H. Iyetomi[a], Y. Ikeda[b], T. Ohnishi[c], H. Takayasu[d] and M. Takayasu[e]

[a]Department of Physics, Niigata University, Niigata 950-2181, Japan
[b]Hitachi Reserch Laboratory, Hitachi Ltd., Ibaraki 319-1221, Japan
[c]Graduate School of law and Politics, University of Tokyo, Tokyo 113-0033, Japan
[d]Sony Computer Science Laboratories, Tokyo 141-0022, Japan
[e]Department of Computational Intelligence and Systems Science, Yokohama 226-8502, Japan

E-mail: *)`iino@ad.sc.niigata-u.ac.jp`

Abstract. We analyze a transaction network of about 800 thousand Japanese firms to elucidate its community structure. Finding community in networks means the appearance of dense connected groups of vertices and sparse connections between groups. We adopt modularity as a quality function of communities introduced by Newman. The modularity optimization is one of effective approaches to find community.

We first use a bottom-up algorithm, which makes the optimization fast by using a greedy algorithm. For the community extraction, the greedy algorithm is widely used, however, may not sufficiently optimize modularity because the optimization tends to be trapped by a local maximum especially for large-scale networks. Alternatively we propose a top-down algorithm with implementation of an annealing method and compare effectiveness of the two algorithms.

We also compare the results of the community analysis with images of network structure visualized by molecular dynamics method. The vertices belonging to the same community are spatially located close to each other. The community structure determined by the modularity optimization is well reproduced in the network structure obtained by molecular dynamics.

1. Introduction

Many complex systems interested by scientific community can be represented as networks. Statistical methods for quantifying and theoretical modeling of networks have been applied with success to extensive networks, including the World Wide Web, citation networks, neural networks, and social network [1, 2]. Real networks are often characterized by communities in which nodes are divided into densely connected groups and on the other hand those groups are sparsely connected. Communities frequently display different statistics from the network as a whole [3]. Extracting communities can also provide coarse-graining view to networks for visualization or other purposes [4]. Community structures are thus helpful for understanding the structure of large complex networks.

We analyze a transaction network consisted of about 800 thousand Japanese firms [1] from

[1] The data used in this paper is collected by TSR and provided by RIETI.

a point of view of community structure and attempt to understand the exhaustive structure of Japanese economy. In the transaction network, each node corresponds to a firm and each link corresponds to a transaction relationship. The transaction network encompasses about all Japanese firms. The reference [5] has analyzed a partial network of manufactures in the same data.

To find a community structure in networks, we need a criterion to evaluate the density of connections of a group for a given partition. Newman proposed modularity as such criterion and extracted community structures by finding the division at the highest modularity value. If a network V is divided into L subsets $\{V_1, V_2, \cdots, V_L\}$ which do not overlap each other and are not empty, the modularity Q is defined as

$$Q = \sum_{i=1}^{L} Q_i = \sum_{i=1}^{L} \left(e_{ii} - a_i^2\right) = \left\{ \frac{1}{2M} \sum_{l \in V_i} \sum_{m \in V_i} A_{lm} - \left(\frac{1}{2M} \sum_{l \in V_i} \sum_{m \in V} A_{lm} \right)^2 \right\}, \tag{1}$$

where A_{lm} is the number of link(s) between node l and node m: an element of adjacent matrix. The symbol a_i denotes the number of links that connect into subset V_i. The symbol e_{ii} represents the link density within subset V_i. When the subsets $\{V_1, V_2, \cdots, V_L\}$ are selected randomly, e_{ii} is canceled out by a_i^2; the term a_i^2 gives the expectation value of the link density in subset V_i. Newman's modularity thus compares the actual link density in a subset with its expectation value. A situation with $Q \simeq 0$ indicates that the network has no statistically significant communities as compared with the randomly connected network and $Q \simeq 1$ corresponds to a network which is partitioned into modules almost perfectly.

The modularity maximization has a fundamental problem; it is NP-complete task [6]. The number of ways to divide a network is given by the sum of the Stirling number of the second kind and so increases super-exponentially with the number of nodes. The true optimization of Q over all possible divisions is very costly. Since the exhaustive search of all possible division would be actually impossible for such a large network as the transaction network with submillion nodes, approximate heuristic methods are usually employed, including greedy agglomeration [7, 8, 9], simulated annealing [10, 11, 12], extremal optimization [13] and spectral methods [3]. In this paper, we employ greedy agglomeration and simulated annealing because of computational performance for large-scale networks.

We explain two methods of finding the community structure briefly in Section 2. The results of the community analysis are given in Section 3. Subsequently, in Section 4, we compare the community structures with the visualized images of the transaction network obtained by molecular dynamics (MD) method [2]. A visualization image gives us an intuitive idea on a structure of large complex networks. Therefore, it is desirable that there is tight relation between the obtained community structure and the visualized image. We report that the community structures are well reproduced in the picture visualized by MD. In Section 5, we elucidate to what extent the attributes (location, industry sector) of firms are reflected in communities of the transaction network. Finally we conclude this paper in Section 6.

2. Method of extracting community

We first use the method due to Clauset *et al.* [8], which makes the optimization fast by using a greedy algorithm. In the initial state of this method, each node belongs to an isolated subset. Every link is calculated the difference ΔQ between the modularity given when both ends of the link are merge to form a new subset and that given by the current subsets. We search the highest ΔQ and merge both ends of the link. This merging process is repeated until there is

[2] ref. Kamehama *et. al.* "Structure analyses of a large-scale transaction network through visualization based on molecular dynamics"

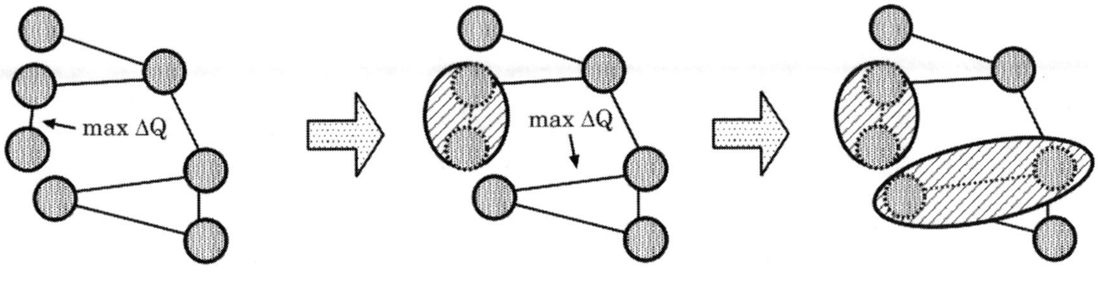

Figure 1. The greedy method proposed by Clauset *et al.* (ref. [8])

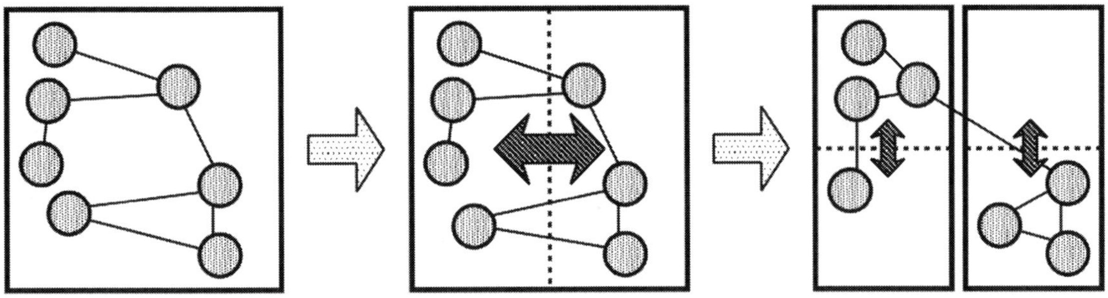

Figure 2. The bisection method worked out here.

no link giveing $\Delta Q > 0$. In this paper, this bottom-up algorithm is called by "greedy method". The outline of the greedy method is shown in Figure 1. Though the greedy method is costless, it may not sufficiently optimize the modularity because greedy optimizations tend to be trapped by a local maximum especially for large-scale systems.

Alternatively, we work out a top-down algorithm that implements the simulated annealing. In the initial state of this method, all nodes in the network belong to one community. We search a dual partitioning that gives the highest modularity value. In the initial state, each node is assigned randomly subset V_A or V_B. Then one node is selected randomly, i.e., belonging to V_A and is tested to transfer it from V_A to V_B. We calculate difference ΔQ between the modularity given when the selected node is transferred and that given by the current subsets. If node l transfers from V_A to V_B, the deference of modularity ΔQ is calculated as

$$\Delta Q = 2\left(e_{lB} - e_{lA}\right) - 2a_l\left(a_B - a_A + a_l\right), \tag{2}$$

where e_{lA} (e_{lB}) denotes the number of links that connect between node l and nodes in subset V_A (V_B) as

$$e_{lA} = \frac{1}{2M}\sum_{m\in V_A} A_{lm}, \tag{3}$$

and a_l denotes the number of links that connect to node l as

$$a_l = \frac{1}{2M}\sum_{m\in V} A_{lm}. \tag{4}$$

The deference ΔQ is yielded from local state around node l. Consequently, this top-down algorithm can be applied to large-scale networks such as the submillion transaction network.

The probability of transition is given by

$$P(\Delta Q, T) = \begin{cases} \exp\left\{\frac{\Delta Q}{T}\right\} & (\text{if } \Delta Q < 0) \\ 1 & (\text{otherwise}) \end{cases}, \quad (5)$$

where T plays a role of temperature. If T is high, the probability of transition tends to become high even if its transition decrease modularity. We repeat transition steps and cool down gradually. As T decreases, the subsets V_A and V_B are fixed. Eventually, we obtain a dual partitioning that gives the highest modularity value. Each of subset V_A and V_B is divided into two parts in the same way. If modularity do not increase any more by splitting any subset, the recursive sequence is terminated. This top-down algorithm is called "bisection method" in this paper. The bisection method is schematically shown in Figure 2.

3. Results

We are now ready to report results of the community structures in the Japanese transaction network. To compare the two optimization methods, we also analyze smaller sample networks. One of the sample networks is power grid network in a western part of US [14]. Nodes in the power grid network represent generators, transformers and substations. Links in the power grconstructionid network represent high-voltage transmission lines. The second sample network is a subset of the transaction network consisting of firms which have employees equal to or more than 500. The subset network can be regarded as the network of large-sized firms in Japanese society because the number of employees is one of the indices of the firm size.

The results of comparison as regards effectiveness of two optimization methods are shown in Table 1. In the results of the power grid network, both methods give almost same values of modularity that exceed 0.9. Furthermore, both methods give almost same numbers of communities.

On the other hand, in the results of the two transaction networks, the bisection method works better. These transaction networks are larger network size than the power grid network. The subset of the transaction network consisting large firms has nearly as many nodes and about four times as many links as the power grid network has. We consider that the greedy method tends to be trapped by a local maximum especially for large-scale networks. The bisection method overcomes this fault.

Next we show the difference of the community size distribution by the two optimization method. Figure 3 represents the cumulative distribution of community size in the power grid network. Difference between the distribution of the greedy method and the bisection method

Table 1. Results of the community analysis on three networks.

	Power Grid	Transaction network	
		(whole)	(number of employees ≥ 500)
number of nodes	4,941	780,544	4,851
number of links	6,594	3,196,282	25,832
greedy method			
number of communities	43	9,540	38
modularity	0.934	0.539	0.504
bisection method			
number of communities	46	1,352	18
modularity	0.936	0.653	0.548

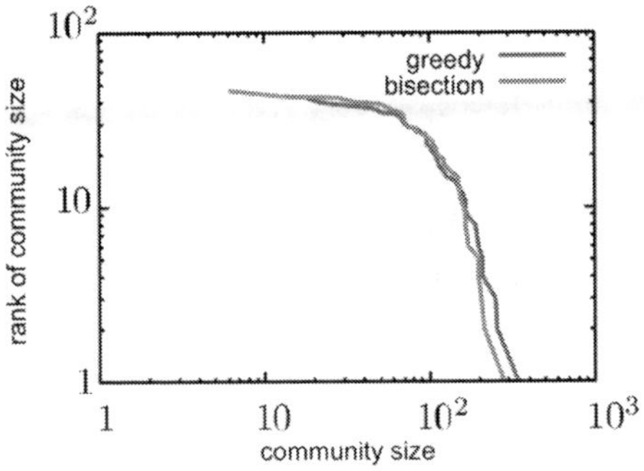

Figure 3. The cumulative distribution of community size in the power grid network.

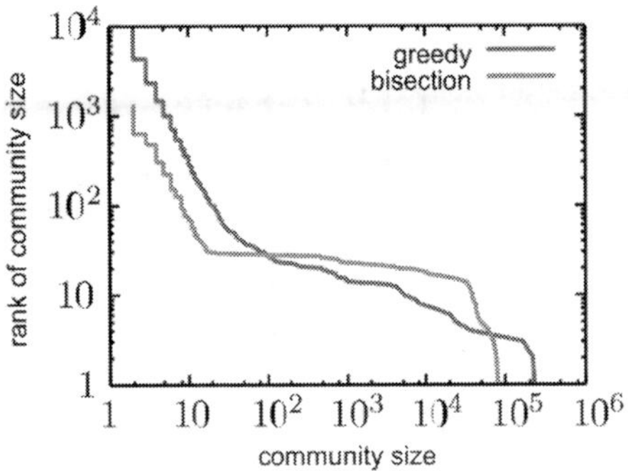

Figure 4. The cumulative distribution of community size in the whole transaction network.

is not detectable. On the other hand, in the whole transaction network, there is considerable difference between the greedy method and the bisection method as shown in Figure 4. The greedy method gives many small communities than the bisection method. Furthermore, the top three communities by the greedy method are exceptionally larger than other communities. The bisection method, however, suppressed such uneven partition of nodes.

In the transaction network data that this paper deal with, the firms has location, industry sector, sales, or other attributes. We characterize communities by checking attributes of firms in each community. Tables 2 and 3 show the fraction of firms' attributes in major communities detected by the two methods.

Table 2. Attributes of firms in major communities detected by the greedy method.

rank	size	prefecture (fraction)	industry sector (fraction)
		Aichi (0.078)	construction (0.593)
1	229,031	Tokyo (0.070)	wholesale/retail trade (0.160)
		Osaka (0.065)	manufacturing (0.147)
		Tokyo (0.190)	wholesale/retail trade (0.459)
2	219,194	Osaka (0.084)	manufacturing (0.252)
		Aichi (0.065)	services (0.098)
		Tokyo (0.153)	wholesale/retail trade (0.262)
3	165,521	Aichi (0.072)	construction (0.249)
		Kanagawa (0.065)	manufacturing (0.227)
		Hokkaido (0.697)	construction (0.438)
4	37,789	Tokyo (0.043)	wholesale/retail trade (0.252)
		Aichi (0.021)	manufacturing (0.109)
		Kanagawa (0.355)	construction (0.684)
5	23,784	Tokyo (0.170)	wholesale/retail trade (0.088)
		Shizuoka (0.154)	real estate (0.085)

Table 3. Same as Table 2, but by the bisection method.

rank	size	prefecture (fraction)	industry sector (fraction)
		Tokyo (0.111)	wholesale/retail trade (0.511)
1	80,134	Hokkaido (0.087)	manufacturing (0.233)
		Aichi (0.055)	eating places/accommodations (0.094)
		Aichi (0.092)	construction (0.702)
2	77,488	Shizuoka (0.091)	wholesale/retail trade (0.106)
		Tokyo (0.090)	manufacturing (0.071)
		Tokyo (0.310)	manufacturing (0.336)
3	72,094	Osaka (0.090)	wholesale/retail trade (0.275)
		Kanagawa (0.081)	information/communications (0.137)
		Tokyo (0.118)	construction (0.628)
4	62,729	Aichi (0.109)	wholesale/retail trade (0.156)
		Kanagawa (0.107)	manufacturing (0.084)
		Fukuoka (0.288)	construction (0.636)
5	47,936	Kagoshima (0.140)	wholesale/retail trade (0.152)
		Kumamoto (0.136)	manufacturing (0.066)

4. Comparison between the community analysis and the MD visualization

In this section, we compare the community structure detected by the modularity with the visualized image by MD method. Figures 5 and 7 distinguish the different communities using different colors. Figures 6 and 8 also show the visualized images, where we discriminate the largest community from others through change of the color. The nodes that belong to a same community tend to be located closely in the visualized image by MD. This result implies that the community structure determined by the modularity optimization is well reproduced in the network structure obtained by MD. This implication is given by both results of the two optimization method.

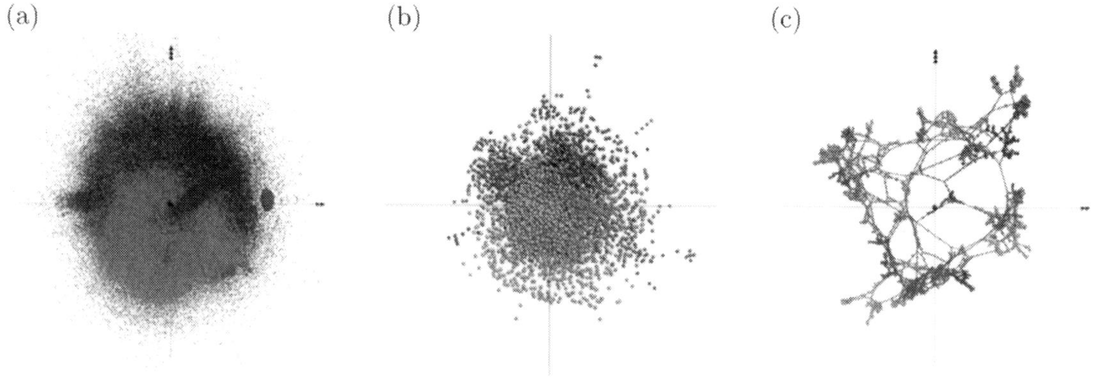

Figure 5. Community structures by the greedy method for (a) the whole transaction network, (b) the partial transaction network with $N_e \geq 500$, (c) the power grid. The base images are the optimized network structures due to MD visualization.

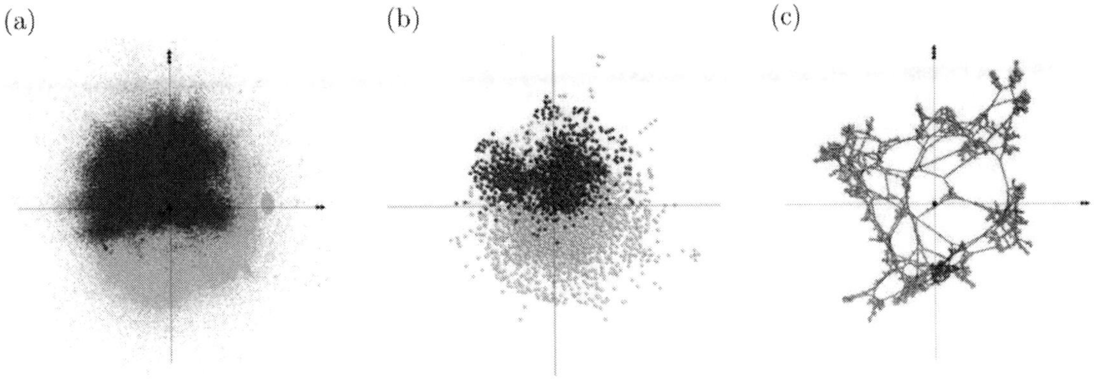

Figure 6. The largest community is illuminated for each of the networks in Figure 5.

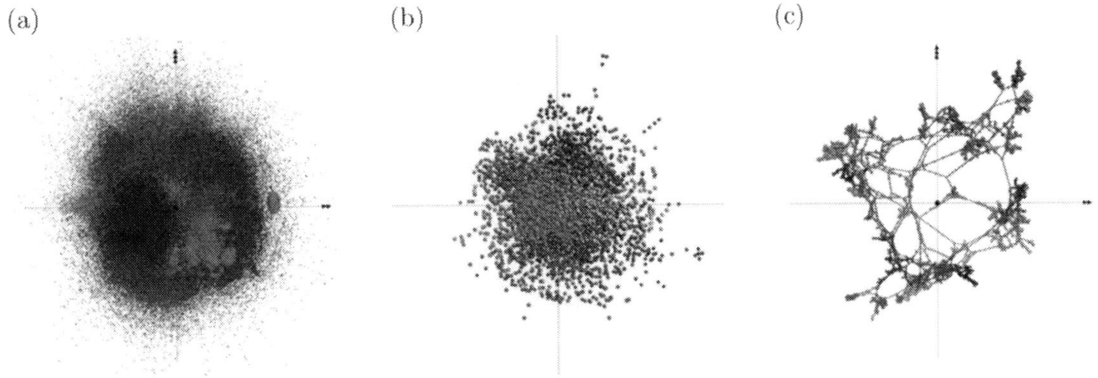

Figure 7. Same as Figure 5, but by the bisection method.

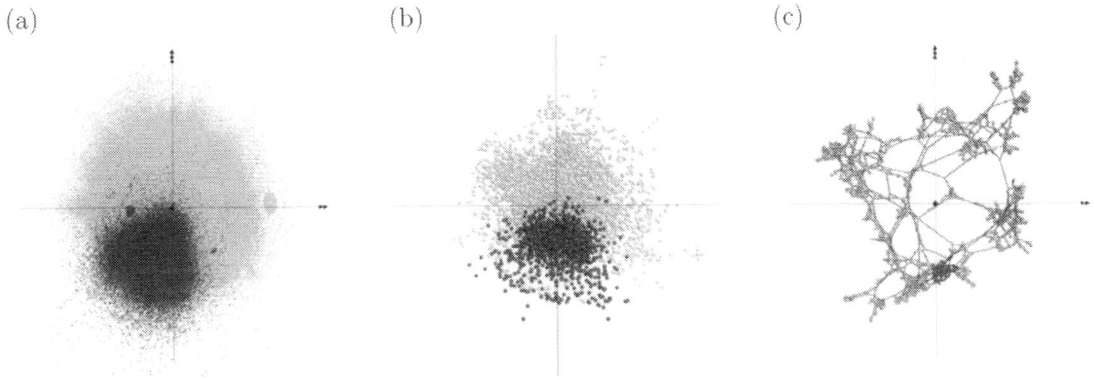

Figure 8. The largest community for each of the networks in Figure 7, coresponding to Figure 6.

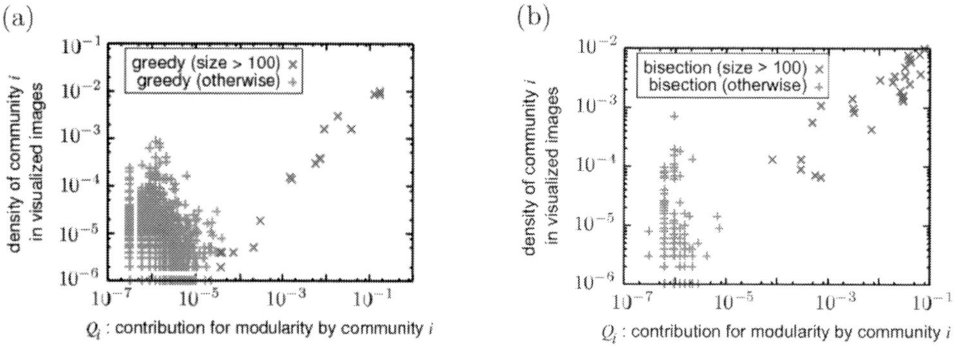

Figure 9. The correlation chart between the contribution for modularity and the density of nodes in visualized images for the whole transaction network: (a) by the greedy method, (b) by the bisection method.

To quantify the good agreement between the community analysis and the MD visualization, we will demonstrate the tendency that highly modularity groups makes high density cluster on visualized images. Figure 9 shows the correlation between the contribution to modularity and the density of nodes in visualized images for the whole transaction network. The contribution to modularity is given by Q_i in Equation (1). The density of nodes in visualized images is defined by

$$D_i = \frac{N_i}{L_i^3}, \quad (6)$$

$$\langle x_i \rangle = \frac{1}{N_i}\sum_{l \in V_i}^{N_i} x_l, \quad \langle y_i \rangle = \frac{1}{N_i}\sum_{l \in V_i}^{N_i} y_l, \quad \langle z_i \rangle = \frac{1}{N_i}\sum_{l \in V_i}^{N_i} z_l, \quad (7)$$

$$L_i = \sum_{l \in V_i}^{N_i} \sqrt{(x_l - \langle x_i \rangle)^2 + (y_l - \langle y_i \rangle)^2 + (z_l - \langle z_i \rangle)^2}, \quad (8)$$

where x_l, y_l and z_l represent the coordinate of node l on visualized images, N_i represents the number of nodes that belong to community i, $\langle x_i \rangle, \langle y_i \rangle, \langle z_i \rangle$ represent center of gravity of community i, and L_i represents the mean distance from the gravity point about nodes that belong to community i. Thus we suppose that L_i^3 is the space occupation volume of community i. The Pearson correlation coefficient between Q_i and D_i by the greedy method is 0.975 and by the bisection method is 0.921. We observe that the communities which give larger contributions to modularity tend to form more strongly-agglutinated clusters in visualized images.

5. Groups classified by attribute of firms

In this section, we investigate whether classification due to the attributes (location, industry sector) of firms gives dense connected groups or not. Figures 10, 11, 12 and 13 show the visualized images color-coded by regions. In the whole transaction network, the nodes that belong to same region tend to be located closely in the MD images as shown in Figures 10 and 11. However, in the subset of the transaction network, the nodes that belong to same region do not tend to be located closely as shown in Figures 12 and 13.

Furthermore, we calculate the modularity of the partition given by industry sectors. The results are shown in Table 4. The modularity for partitioning according to industry sectors should be lower than the optimized modularity. Figures 14 and 15 show the visualization color-coded by industry sectors that nodes belong to. The firms of construction and finance/insurance are located closely, however, the firms of real estate and some other industry sectors are dispersed as shown in Figures 15.

(a) (b) (c)

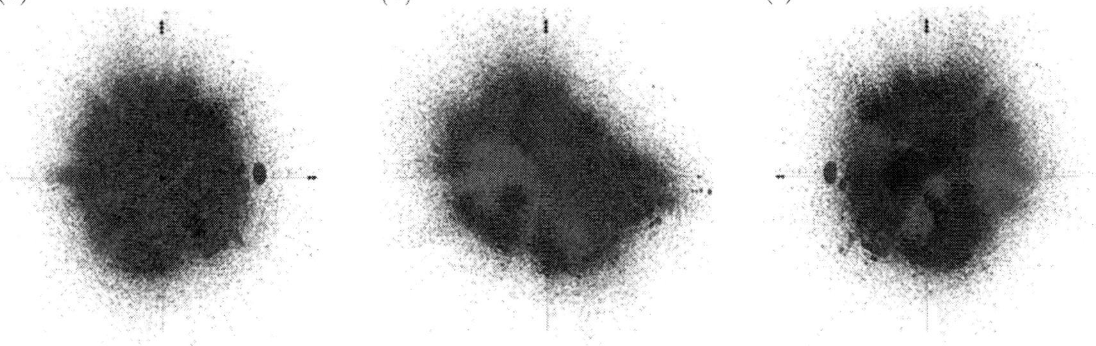

Figure 10. The MD images of the whole transaction network; color-coded by all prefectures, which are drawn from different view angles.

(a) (b) (c)

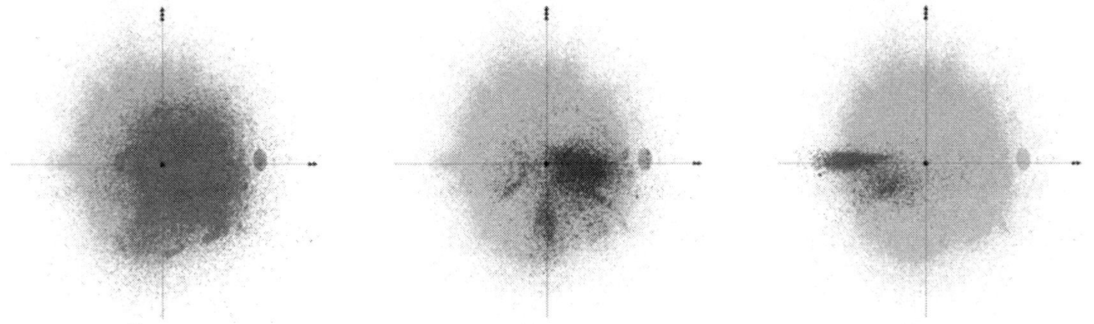

Figure 11. The MD images of the whole transaction network: in which (a) Tokyo, (b) Niigata, and (c) Okinawa are illuminated separately.

(a) (b) (c)

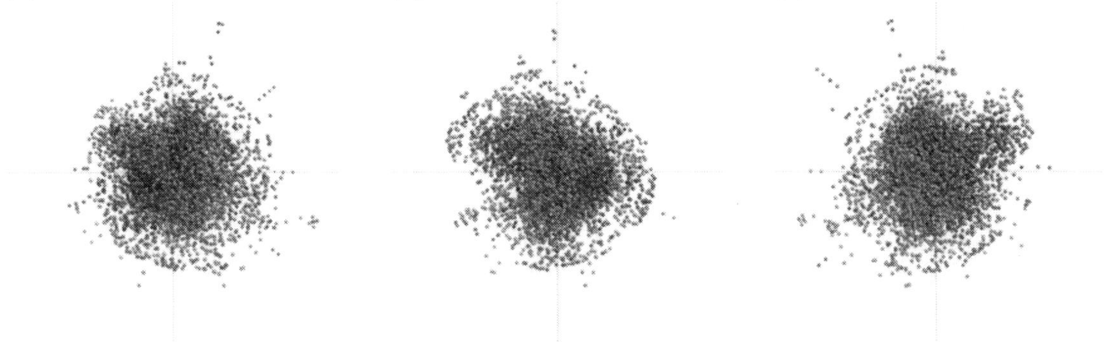

Figure 12. The MD images of the subset of the transaction network; color-coded by all prefectures, which are drawn from different view angles.

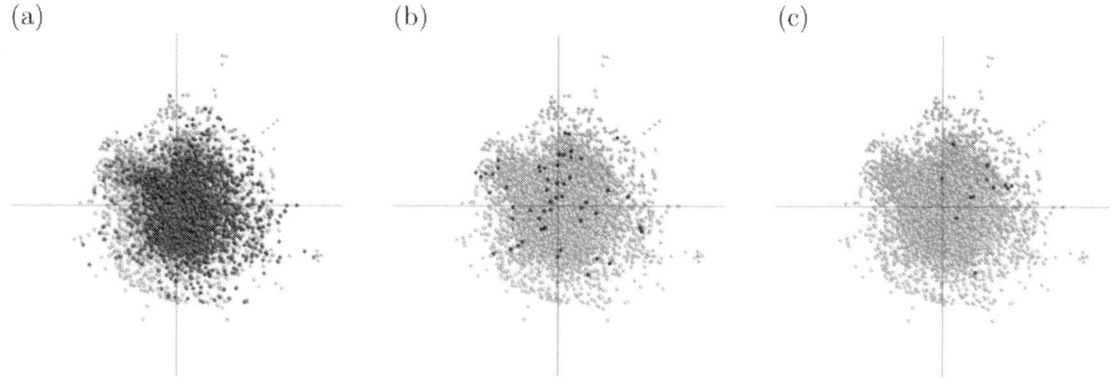

Figure 13. The MD images of the subset of the transaction network: in which (a) Tokyo, (b) Niigata, and (c) Okinawa are illuminated separately.

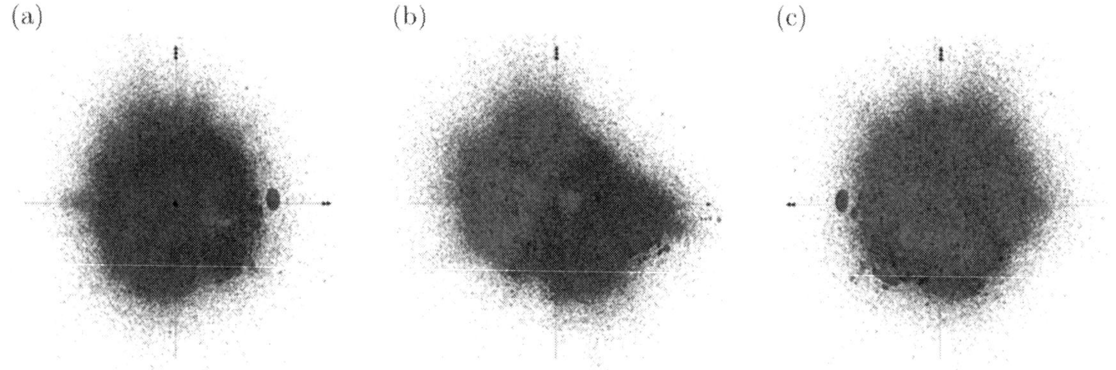

Figure 14. The MD images of the whole transaction network; color-coded by industry sectors, which are drawn from different view angles.

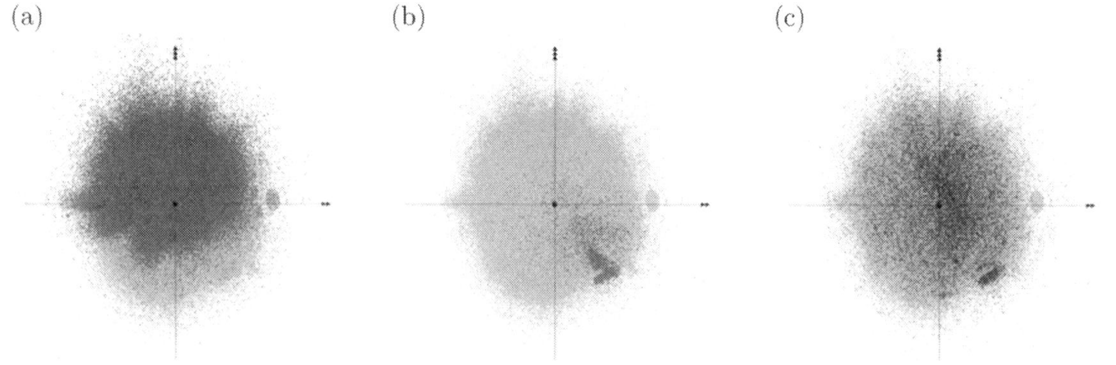

Figure 15. The MD images of the whole transaction network: in which (a) construction, (b) finance/insurance, and (c) real estate are illuminated separately.

Table 4. Modularity of the partition given by attribute.

	Transaction network	
	(whole)	(number of employees ≥ 500)
prefectures		
number of groups	47	47
modularity	0.483	0.109
industry sectors		
number of groups	18	18
modularity	0.175	0.134

We calculate the modularity of the division by prefectures and confirm whether the firms which belong to same area have dense connections. The results are shown in Table 4. In the whole transaction network, the modularity of the division by prefectures is 0.483. The transaction network has highly-modularized groups that contain geographically close firms because the modularity value 0.483 is close to 0.539: the modularity of the community determined by the greedy method.

On the other hand, in the subset of the transaction network consisting large firms, the modularity of the division by prefectures is 0.109; this value is remarkably less than the modularity of the community determined by the greedy method, 0.504. Thus the large firms that employ many people tend to have nation-wide connection. We conclude the difference between Figures 11 and 13 is caused by the difference of the modularity. If a partition gives a high modularity, the MD image shows the partition as cluster. If a partition gives a low modularity, the visualization shows dispersed color configuration. The MD image is helpful to obtain an outline of the community structure.

We investigated the tendency that highly modularity groups makes high density cluster on visualized images as previous section. Figure 16 shows the correlation between the contribution for modularity Q_i and the density of nodes D_i in MD images for the whole transaction network. The Pearson correlation coefficient between Q_i and D_i by the prefectures is 0.976 and by the industry sectors is 0.880. The groups which give large contribution to modularity tend to make strongly agglutinated clusters in visualized images, in the same way as the community structure extracted. The firms of construction and finance/insurance are located closely on the visualized

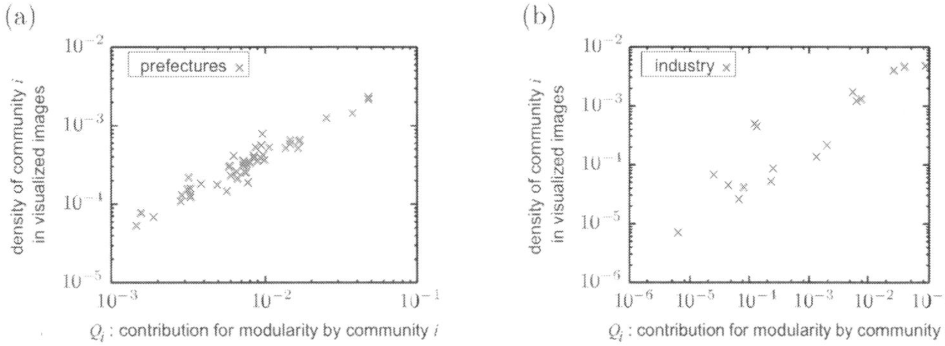

Figure 16. The correlation between the contribution for modularity and the density of nodes in MD images for the whole transaction network: (a) by the prefectures, (b) by the industry sectors.

images because they give large contribution to modularity. However, some other groups of nodes belonging to same industry sectors give lower contribution to modularity compared to the result of prefectures. Eventually, the total modularity obtained by summing the partial contributions through all industry sectors is lower than prefectures.

6. Conclusion

We analyze the transaction network of submillion Japanese firms by detecting its community structure. For large networks, the method of community detection is restricted to costless algorithm. We compare the two optimizing method because the optimization may be trapped by a local maximum. As a result, the bisection method works better than the greedy method especially on large-scale networks as the nation-wide transaction network. The community structure obtained by the modularity optimization is reflected in the network structure given by molecular dynamics. A partition that has high modularity is visualized as a cluster. We confirm that the community structure is shown as clusters by the color-coded images.

In the visualized images of the whole transaction network by the molecular dynamics, the firms in a same prefecture or region tend to be located closely. Thus we elucidate the modularity of the partition by each firm's address and confirm that the modularity is close to the value obtained by the optimizing method. The transactions between firms tend to be confined in same prefecture or region. On the other hand, in the visualized images of the subset of the transaction network consisting large firms, the firms in same prefecture or region tend to disperse. The modularity of the partition by each firm's address in the subset of the transaction network is far from the value obtained by the optimizing method. Large firms are connected through transaction over local boundary. The visualization by molecular dynamics is helpful to illustrate the community structures of large-scale complex networks.

Acknowledge

We thank the Research Institute of Economy, Trade and Industry (RIETI) for having provided us with the data on transaction relationship between firms in Japan. It was critical for us to carry out the present study.

References

[1] R. Albert and A.-L. Barabási, *Rev. Mod. Phys.* **74** (2002), 47.
[2] M. E. J. Newman, *SIAM Rev.* **45**, 167 (2003).
[3] M. E. J. Newman, *PNAS* **103** (2006), 8577.
[4] M. E. J. Newman and M. Girvan, *Phys. Rev.* E **69** (2004), 026113.
[5] Y. Fujiwara and H. Aoyama, 0806.4280[e-print arXiv].
[6] U. Brandes, D. Delling, M. Gaertler, R. Görke, M. Hoefer, Z. Nikoloski, and D. Wagner, *IEEE Transactions on Knowledge and Data Engineering* **20** (2008), 172.
[7] M. E. J. Newman, *Phys. Rev.* E **69** (2004), 066133.
[8] A. Clauset, M. E. J. Newman and C. Moore, *Phys. Rev.* E **70** (2004), 066111.
[9] V. D. Blondel, J.-L. Guillaume, R. Lambiotte and E. Lefebvre, *J. Stat. Mech. p.* P10008 (2008).
[10] R. Guimerà and L. A. N. Amaral, *Nature* **433** (2005), 895.
[11] A. Medus, G. Acuña and C. O. Dorso, *Physica* A **358** (2005), 593.
[12] J. Reichardt and S. Bornholdt, *Phys. Rev.* E **74** (2006), 016110.
[13] J. Duch and A. Arenas, *Phys. Rev.* E **72** (2005), 027104.
[14] D. J. Watts and S. H. Strogatz, *Nature*, **393** (1998), 440.

Structure analyses of a large-scale transaction network through visualization based on molecular dynamics

K Kamehama[1], T Iino[1], H Iyetomi[1], Y Ikeda[2], T Ohnishi[3], H Takayasu[4] and M Takayasu[5]

[1] Department of Physics, Niigata University, Niigata 950-2181, Japan
[2] Hitachi Research Laboratory, Hitachi Ltd., Ibaraki 319-1221, Japan
[3] Graduate School of Law and Politics, the University of Tokyo, Tokyo 113-0033, Japan
[4] Sony Computer Science Laboratory, Tokyo 141-0022, Japan
[5] Department of Computational Intelligence and Systems Science, Tokyo Institute of Technology, Yokohama 226-8502, Japan

E-mail: hiyetomi@sc.niigata-u.ac.jp

Abstract. We shed light on industrial structure of the economic system in Japan by combining visualization technique and community analysis in this and an accompanying paper. Here we focus on visualization of a production network comprising submillion nodes (firms) and three million links (transaction relations). A network structure is optimized through molecular dynamics simulation with a spring-electrical model in a three-dimensional space. The lowest-energy state in the model, corresponding to a crystalline state in a physical system, is expected to give a comprehensible view on the network. Then we discuss how firms are distributed in the optimized network structure by classifying them according to sectors, sizes or regions. The distributions of firms reflect characteristics of individual classifications. Also propagation of the shock due to a recent economic scandal over the network is visualized.

1. Introduction

A wide variety of systems as encountered in informational, social, economical, and biological sciences are recognized as complex networks consisting of an enormous number of linked elements [1, 2, 3, 4]. As an example, firms form a complex network through their business transactions for their production activities. A firm obtains materials from suppliers (upstream firms) and sell its products to customers (downstream firms). The business relationship so established between firms may also reflect how close are they geographically, financially, politically, historically, mentally and so on. Fortunately we can have access to an empirical database on production network in which one million firms with four million links in Japan are listed. It virtually encompasses the whole production activities over the nation.

The objective of the present work is to develop a visualization method adapted for drawing such a large-scale network and to apply the method so obtained to the empirical data. Complex networks are highly difficult to analyze all connections because of their multiplicity and complexity. Visualization is one of useful tools to illuminate structural properties of networks.

Of course visualization is not good at everything, but it helps us a lot to delve into complicated structures by stimulating our intuition.

Various algorithms have been developed to visualize networks whose structural elements are nodes (firms) and links (transaction relations between firms) [5]. Here we adopt a spring-electrical model in which pairs of firms with direct transaction relations are physically connected with springs and any pairs of firms repel each other through a repulsive Coulomb force [6]. The attractive force due to the spring keeps intimate firms close in a space. On the other hand, the repulsive Coulomb force tends to distribute firms uniformly over the available space and to prevent entanglement of the transaction network. We then take full advantage of a molecular dynamics (MD) method [7, 8] for an optimized configuration of nodes in the model. The ground state in the model is a leading candidate for it. The MD simulation works well to reproduce an ordered structure with the lowest energies such as crystals of materials through slow cooling starting from any initial configurations. We expect it is also successful in visualizing the network.

In the accompanying paper [9] we also devote ourselves to development of a method for extracting community structures inherent in such a nationwide economic network. Finding communities in a network is to separate densely connected groups of nodes from the remainder. Then the network can be regarded as consisting of those dense groups sparsely connected to each other. The structural heterogeneity may play an important role in determining dynamics of macroeconomy such as business fluctuations, one of long-standing problems in the economic theory [10]. Maximization of the modularity, a quality function for community detection proposed by Newman and Girvan [11], is one of promising approaches to identifying community structures involved in complex networks [12, 13, 14, 15, 16, 17, 18, 19]. We thus demonstrates that combination of the two approaches, visualization and community analysis, to the transaction network possesses potential utility in elucidating the economic structure in Japan.

This paper is organized as follows. In the next section we briefly review the dataset used here. Section 3 explains the MD visualization method for an optimized network structure. In Sec. 4 we show the optimized network structure and give examples of analyses using visualized images based on it. Section 5 summarizes the present results with concluding remarks.

2. Transaction Relations between Firms

We use a set of data on transactions between firms in Japan, compiled by Tokyo Shoko Research, Inc. It is almost the same dataset as analyzed in Ref. [20], where details of the data are described including how the research agency collected information. Here we just mention the dataset contains information in 2004 on individual firms exceeding one million in Japan and more than four million transactions between firms. The total number of firms in Japan is now estimated to be 2.5 million by the census due to the Ministry of Internal Affairs and Communications. By removing data irrelevant to the production network (isolated firms, incomplete data, etc.), we reduced the original dataset to the one consisting of $N = 780,544$ nodes (firms) and $M = 3,196,282$ links (transaction relations).

It has already been established that the connections between firms have characteristics of a scale-free network [20, 21]. Namely, the number of links (degree) originating from a node has a power-law distribution; there exists firms acting as hubs with a number of connections at every scale. The scale-free networks are observed in various places, including the topology of web-page links, the collaborative network of Hollywood actors, the power grid of U.S., and so on [1, 2, 3, 4]. Preferential attachment is a possible mechanism for formation of the scale-free networks; probability of linking to a node is proportional to the number of existing links that node has.

3. Molecular Dynamics Visualization

We carried out MD simulation for an optimized structure of the transaction network in a three-dimensional space. The interaction potential between nodes for the present simulation is explicitly written as

$$\Phi(r_{ij}) = \frac{Q_i Q_j}{r_{ij}} + \frac{1}{2} k_{ij} r_{ij}^2, \quad (1)$$

where Q_i is a Coulomb charge for node i and k_{ij} is a spring constant between nodes i and j which are directly connected. Here we neglect the direction of links (flow of goods or money), and assume that Q_i and k_{ij} take identical values for every node and every pair, respectively. We used the velocity Velret algorithm [7, 8] for integrating the equation of motion for a system of classical particles interacting through the potential (1).

The MD simulation on a system consisting of charged particles of million order is not a simple task because of the long-range nature of the Coulomb force; the computational task is order of N^2 without any tricks. To speed up the computation, we thereby implemented the hierarchical tree algorithm due to Barnes and Hut [22, 23] into our MD code. The whole space is divided by cells at hierarchical levels. The interactions of a given particle with other remote particles are replaced by the interactions with the cells to which those particles belong. As the distance between particles is larger, cells at higher level are used. On the other hand, the force on the particle is calculated directly for its neighbors residing in the same cell or the adjacent cells at the lowest level. This trick reduces the computational task to $O(N \log N)$.

The nodes were distributed randomly in a cubic box at the initial time. The given configuration immediately led to a very high temperature state in which all nodes were moving almost freely. Then we gradually decreased temperature of the system through reduction of the kinetic energies of nodes to reach a zero-temperature state. Relative magnitudes of the potential parameters control the final structure obtained, although their absolute magnitudes are totally irrelevant. We confirmed that the final result was essentially independent of initial configurations and cooling processes adopted. We show an example of the relaxation process in figure 1.

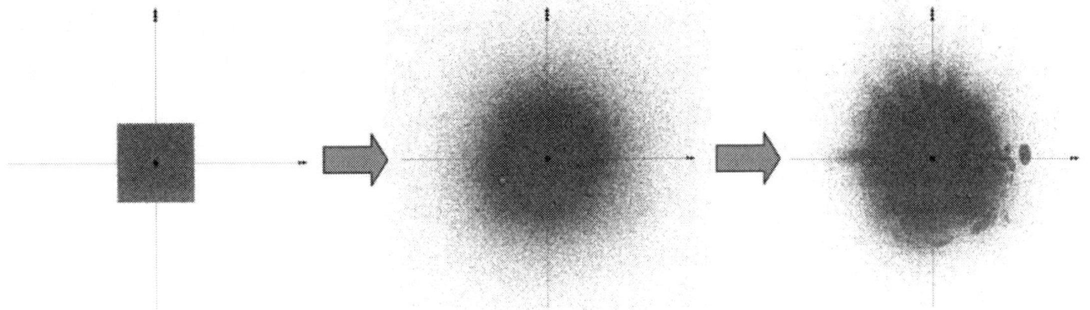

Figure 1. Relaxation process in the MD simulation, where only nodes are drawn; initial state (left), intermediate state (middle) and final state with zero temperature (right).

4. Visualized Results and Discussion

The optimized network structure as shown by the rightmost drawing in figure 1 will be a basis for the present study, and will be utilized in the community analysis in the accompanying paper [9].

We first decompose firms into eight large sectors of industry: Construction, Manufacturing, Services, Real Estate, Transport, Wholesales/Retail Trade, Information/Communications and

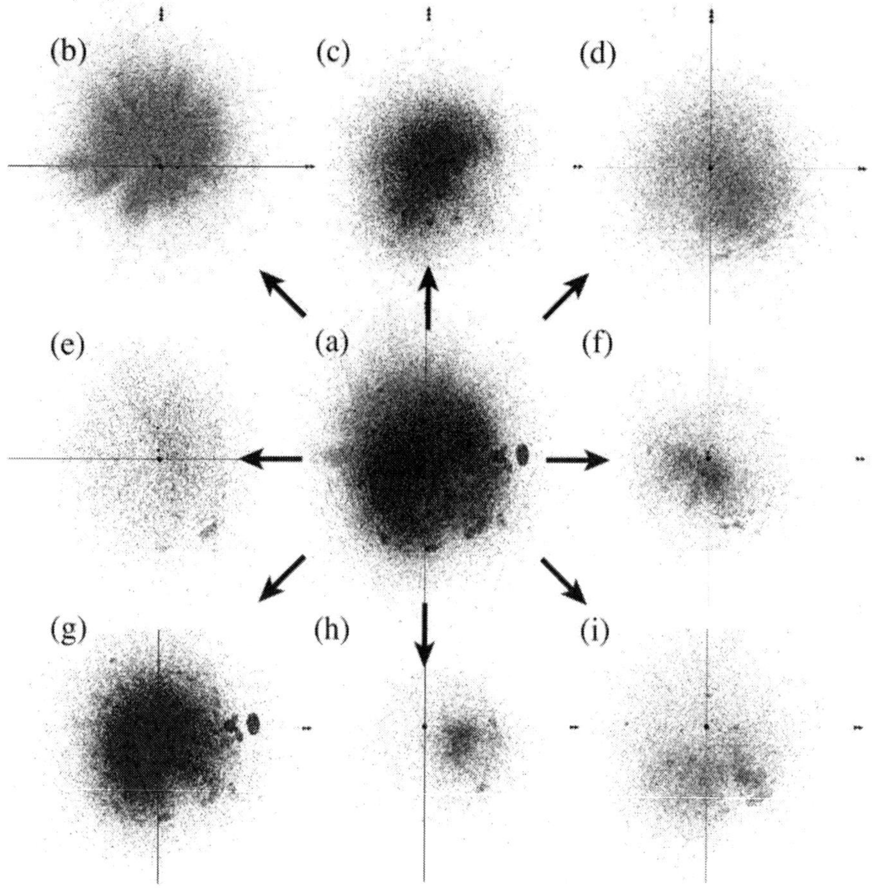

Figure 2. The optimized network structure of the whole economy which are decomposed into eight sectors designated by different colors (a); the central view is identical to the right drawing in figure 1. The sectors are listed as follows: Construction (b), Manufacturing (c), Services (d), Real Estate (e), Transport (f), Wholesales/Retail Trade (g), Information/Communications (h), Others (i).

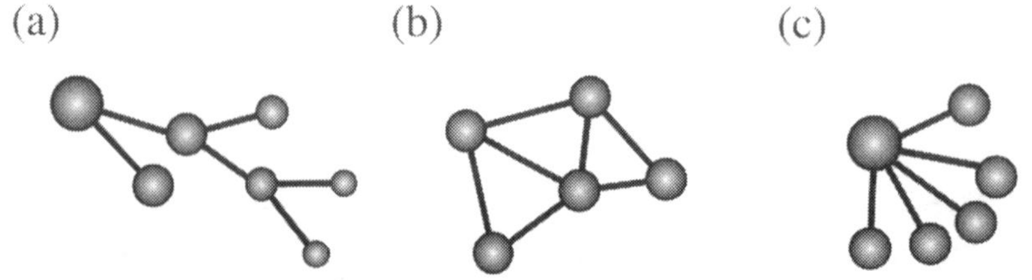

Figure 3. Schematic drawing of typical transaction structures in three sectors, construction (a), manufacturing (b), and wholesales/retail trade (c).

Table 1. Properties of partial networks within sectors and those of the whole network. The symbols A through F stand for the following quantities in each sector: A, fraction of nodes in percentage; B, power-law exponent [4] for the degree distribution; C, average degree of nodes; D, fraction of intralinks as compared with total links; E, average path length; F, cluster coefficient [4].

Sector	A	B	C	D	E	F
Construction	32.7	2.56	4.34	42.3	5.78	0.0379
Manufacturing	18.0	2.50	5.76	25.5	4.99	0.0484
Information/ Communications	2.07	2.62	3.95	19.0	4.96	0.0435
Transport	3.43	2.62	3.70	17.0	4.78	0.0502
Wholesales/ Retail trade	28.0	2.51	5.20	26.8	5.16	0.0365
Real estate	2.40	2.75	2.17	6.77	5.68	0.0172
Services	8.66	2.55	2.42	9.36	7.18	0.0250
Others	4.77	2.37	2.58	6.81	7.04	0.0227
Whole	—	2.50	8.20	—	4.77	0.0447

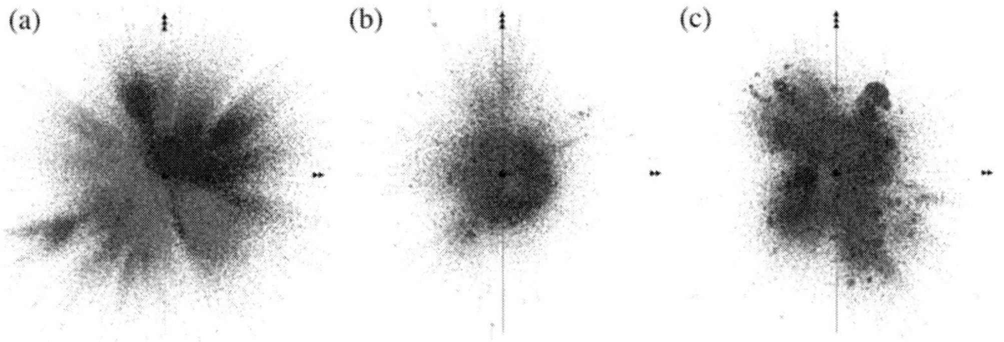

Figure 4. Drawings of partial networks for three sectors: Construction (a), Manufacturing (b), and Wholesales/Retail Trade (c). The networks are colored to distinguish regions (see figure 5 below)

Others. Among them, Construction, Manufacturing and Wholesales/Retail Trade are big three sectors covering almost 80% of total firms. Figure 2 shows how firms of each sector are distributed in the network and gives us an idea on mutual positions of the sectors in the economic system. Roughly speaking, the major sectors are placed as follows: firms in Wholesales/Retail Trade gather in the front side of the space in figure 2; most of firms in Manufacturing are positioned in the vicinity of the center in the space; firms in Construction occupy the back side of the space. Firms belonging to Services and Real Estate are spread over the entire space. In contrast, firms in Transport and Information/Communications form a core at the center. This figure thus constitutes a kind of industrial map in Japan.

Also we see that each sector has its own structural characteristics. For instance, the construction sector has brush-like structures. Such a feature indicates that hierarchical relationship (contractors, sub-contractors, sub-sub contractors and so on) prevails in the sector. Firms in the manufacturing sector, in contrast, are tightly connected to each other, forming a dense cluster. The wholesales/retail trade sector is characterized by a number of mushroom-like structures of various sizes. This demonstrates relationship between a wholesaler and many retailers associated with it are popular at every trade scale in Japan. The typical structures of those three sectors are illustrated in figure 3.

Properties of partial networks within individual sectors are listed in table 1; the partial networks also possess the scale-free nature. And three of them corresponding to the major sectors are depicted in figure 4. Those network structures were optimized by separate MD runs. The characteristic properties (recall figure 3) that we have remarked as regards the three sectors are illuminated more clearly here. We note the partial network of the manufacturing sector was visualized in advance by Fujiwara and Aoyama [20].

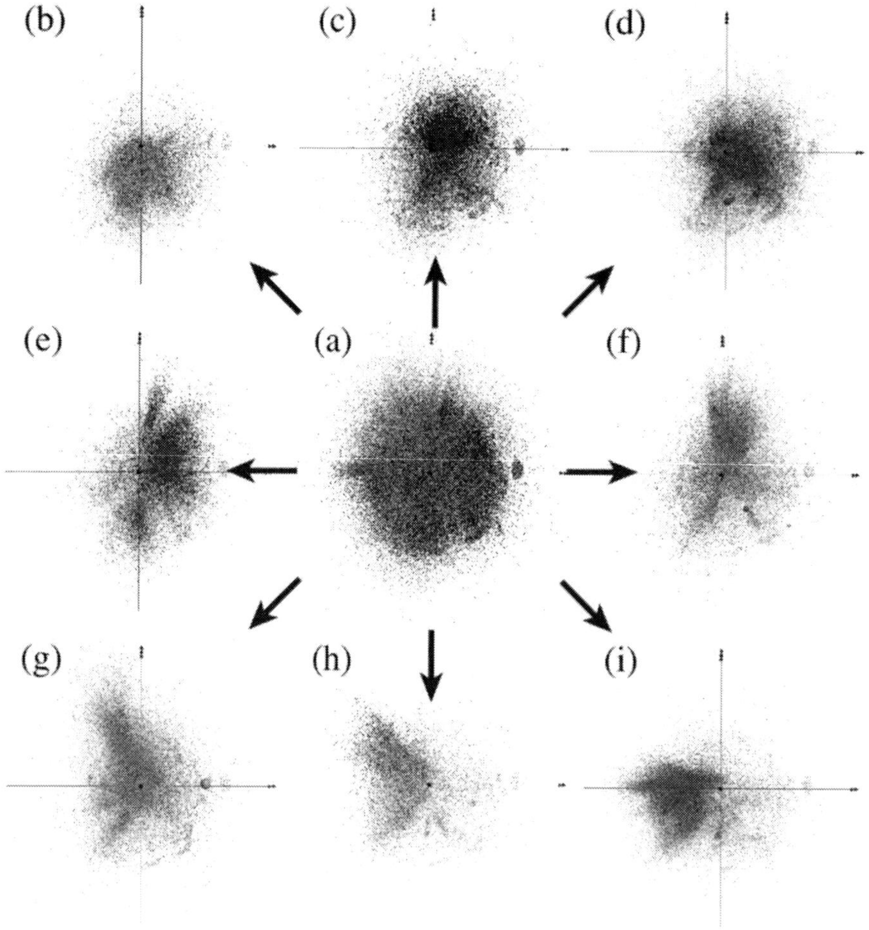

Figure 5. The same network structure as has been shown in figure 2, but classified according to regional decomposition (a). The 47 prefectures in Japan are grouped into eight regions: (b) Hokkaido & Tohoku (Aomori, Iwate, Akita, Miyagi, Yamagata and Fukushima), (b) Kanto excluding Tokyo (Tochigi, Ibaraki, Gunma, Saitama, Chiba and Kanagawa), (c) Tokyo, (d) Central East (Niigata, Toyama, Nagano, Yamanashi and Shizuoka), (f) Central West (Fukui, Ishikawa, Gifu, Aichi, Mie), (g) Kinki (Shiga, Kyoto, Osaka, Hyogo, Nara, Wakayama), (h) Chugoku (Shimane, Okayama, Tottori, Hiroshima and Yamaguchi) & Shikoku (Kagawa, Tokushima, Ehime and Kochi) (i) Kyushu (Fukuoka, Saga, Nagasaki, Oita, Kumamoto, Miyazaki and Kagoshima) & Okinawa.

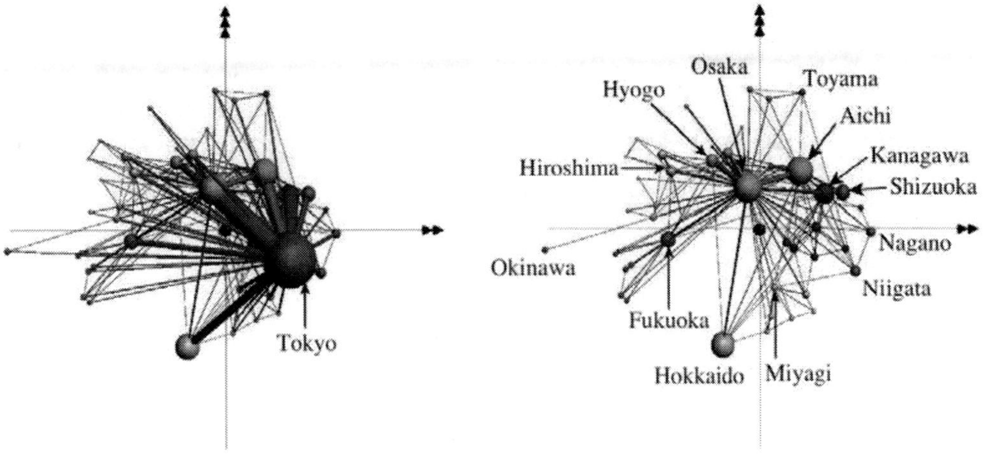

Figure 6. Coarse-grained pictures of the optimized network as shown in Fig 5 at prefecture level with (left) or without (right) Tokyo, where some of the prefectures are designated. The size of spheres representing the prefectures reflects how many firms each of them has.

So far we have paid our attention to the role of the industrial sectors in the optimized network. Alternatively, we classify firms according to which regions they belong to. The result is shown in figure 5, where the 47 prefectures in Japan are grouped into eight regions: Hokkaido & Tohoku, Kanto excluding Tokyo, Tokyo, Central East, Central West, Kinki, Chugoku & Shikoku and Kyushu & Okinawa. The left picture in figure 6 shows a coarse-grained network at prefecture level which was derived from the optimized network. Each node is placed at the center of gravity of firms belonging to the prefecture it represents. Since Tokyo occupies a very peculiar position linking over the whole nation, we also prepared a see-through picture, the right one in the figure, by discarding Tokyo. The nodes representing the prefectures are positioned in a similar way to the corresponding geographical location from north (Hokkaido) to south (Okinawa). In passing we note that the construction sector is well separated regionally as seen in figure 4(a).

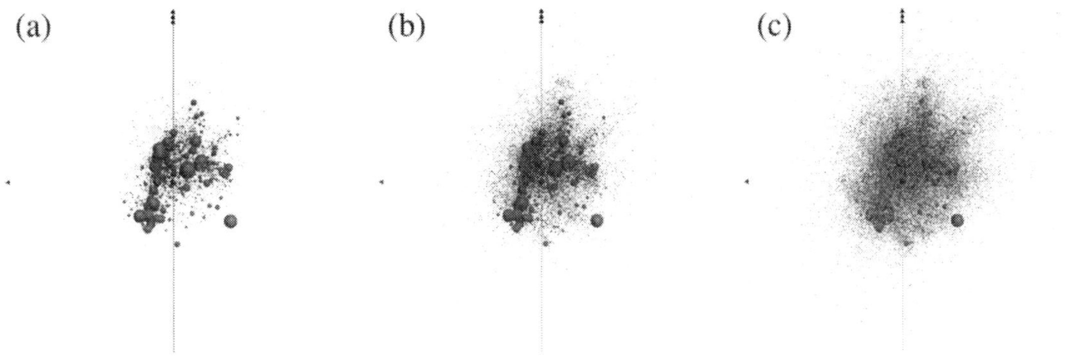

Figure 7. Partial drawings of the optimized network structure in figure 1 where firms are selected according to their size. The size of firms is measured by the total number N_e of employees: (a) for $N_e \geq 500$, (b) for $N_e \geq 100$, and (c) for $N_e \geq 25$. The radius of each sphere representing a firm scales with the number of links it has.

Figure 7 shows how firms with different sizes are distributed in the optimized network, where the size of a firm is represented by its number of employees. We see that larger firms prefer to reside at the center. Recalling that the present visualization method tends to locate nodes of large degree around the center, we calculated the correlation coefficient r between the size and the degree of firms. Certainly the size of firms strongly correlates with their degree as confirmed by the result $r = 0.65$.

Finally figure 8 depicts how a shock due to a scandal over sale of contaminated rice by a food distribution company propagate through the network. The scandal was brought to light in fall, 2008. It was revealed that the distributor illegally sold contaminated rice for human consumption; such rice was only allowed for industrial use. The distributor sold the rice to wholesalers, who sold it to food makers down the production line. As shown in figure 8, the shock is directed to the center of the network indicating amplified impact on the economic system. Actually a number of downstream firms and consumers were affected.

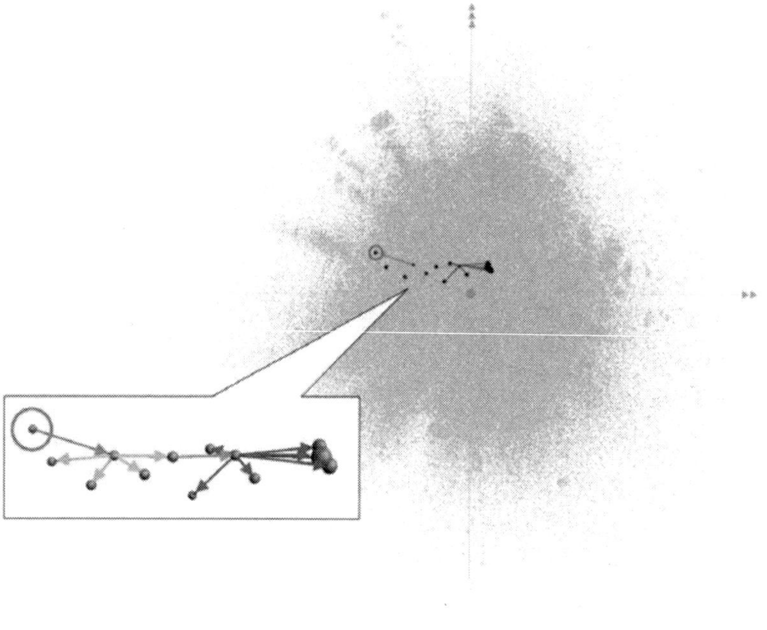

Figure 8. Propagation of a shock due to the scandal over sale of contaminated rice by a food distribution company through the production network. The red circle depicts the trigger company.

5. Concluding Remarks

We optimized the network structure arising from three million transactions between submillion firms using MD simulation, and visualized the network in a three-dimensional space. We then analyzed how firms were distributed in the optimized network from the point of view of sectors, sizes or regions. We thus succeeded in elucidating characteristics of individual classifications through the distributions of firms. Also we ascertained the gravity of the recent economic scandal by displaying propagation of the shock over the network. This paper has just initiated possible applications of the visualization method developed here to analyzing the transaction network. We hope it will serve as a useful tool to study the industrial structure and stability of the economic system in Japan. The accompanying paper will proceed one step further in this direction by emphasizing community detection.

Acknowledgments
We deeply thank the Research Institute of Economy, Trade and Industry (RIETI) for having provided us with the data on transaction relations between firms in Japan. We could not have carried out the present work without the offer.

References

[1] Watts D J 1999 *Small Worlds: The Dynamics of Networks Between Order and Randomness* (Princeton: Princeton University Press)
[2] Barabási A-L 2002 *Linked: The New Science of Network* (Cambridge: Perseus Publishing)
[3] Buchanan M 2002 *NEXUS: Small Worlds and the Groundbreaking Science of Networks* (New York: W W Norton&Co)
[4] Newman M E J 2003 *SIAM Review* **45** 167
[5] Di Battista G, Eades P, Tamassia R and Tollis IG 1998 *Graph Drawing: Algorithms for the Visualization of Graphs* (Englewood Cliffs, NJ: Prentice Hall)
[6] Hu Y 2006 *The Mathematica Journal* **10** 37 and references therein.
[7] Allen M P and Tildesley D J 1987 *Computer Simulation of Liquids* (Oxford: Oxford University Press)
[8] Frenkel D and Smit B 2002 *Understanding Molecular Simululation: From Algorithms to Applications* 2nd ed (San Diego: Academic Press)
[9] Iino T, Kamehama K, Iyetomi H, Ikeda Y, Ohnishi T, Takayasu T and Takayasu M, a contribution in this volume.
[10] Aoki M and Yoshikawa H 2007 *Reconstructing Macroeconomics: A Perspective from Statistical Physics and Combinatorial Stochastic Processes* (New York: Cambridge University Press)
[11] Newman M E J and Girvan M 2004 *Phys. Rev.* E **69** 026113
[12] Newman M E J 2004 *Phys. Rev.* E **69** 066113
[13] Clauset A, Newman M E J and Moore C 2004 *Phys. Rev.* E **70** 066111
[14] Guimerà R and Amaral L A N 2005 *Nature* **433** 895
[15] Duch J and Arenas A 2005 *Phys. Rev.* E **72** 027104
[16] Danon L, Duch J, Arenas A and Diaz-Guilera A 2005 *J. Stat. Mech.: Theory Exp.* P09008
[17] Newman M E J 2006 *PNAS* **103** 8577
[18] Sales-Pardo M, Guimerà R, Moreira A A, Amaral L A N 2007 *PNAS* **104** 15224
[19] Blondel V D, Guillaume J-L, Lambiotte R and Lefebvre E 2008 *J. Stat. Mech.: Theory Exp.* P10008
[20] Fujiwara Y and Aoyama H 0806.4280[e-print arXiv]
[21] Fujiwara Y, Aoyama H and Souma W 2006 *Econophysics and Sociophysics*, ed B K Chakrabarti, A Chakraborti and A Chatterjee (Berlin: Wiley-VCH) p 99
[22] Barnes J and Hut P 1986 *Nature* **324** 446
[23] Pfalzner S 1996 *Many-Body Tree Methods in Physics* (Cambridge University Press, New York)

The low volatility fluctuations regime of the exponential Ornstein - Uhlenbeck model

G Bormetti[1,2], V Cazzola[1,3,2], D Delpini[3,2], G Montagna[3,2,1] and O Nicrosini[2,1]

[1] Istituto Universitario di Studi Superiori, Centro Studi Rischio e Sicurezza, Viale Lungo Ticino Sforza, 56 27100 Pavia, Italy

[2] INFN, Sezione di Pavia, Via A.Bassi, 6 27100 Pavia, Italy

[3] Università degli Studi di Pavia, Dipartimento di Fisica Nucleare e Teorica, Via A.Bassi, 6 27100 Pavia, Italy

E-mail: giacomo.bormetti@pv.infn.it; valentina.cazzola@pv.infn.it; danilo.delpini@pv.infn.it; guido.montagna@pv.infn.it; oreste.nicrosini@pv.infn.it

Abstract. In this work the analytical characterization of the probability density of financial returns in the exponential Ornstein-Uhlenbeck model is addressed. Since prices are driven by a Geometric Brownian motion, whose diffusion coefficient is expressed through an exponential function of the mean-reverting process Y, an exact expression for the density is hard to be drawn. However, we derive the closed-form formula for the characteristic function and for the cumulants of the approximating linear model under the low volatility fluctuations regime. Theoretical results are confirmed numerically by use of Monte Carlo simulations. The effectiveness of the analytical predictions is tested on the German DAX30 Index, finding a good agreement between the empirical data and the theoretical description.

1. Introduction

Since the pioneering work of Bachelier [1] diffusion processes have been natural candidates for the modeling of the stochastic evolution of financial quantities. The assumption of a Geometric Brownian motion as the dynamics driving prices fluctuations led to the famous Black, Scholes and Merton theory of rational option pricing. In their framework the volatility is assumed to be constant, however it has been demonstrated empirically that it fluctuates along time and its stochastic behavior is responsible for the excess of kurtosis observed in the empirical returns distributions. In financial literature the idea of modeling the volatility as a random variable as been widely exploited [2–6]. The research in this field has also been strongly enhanced by the works of Carr and Madan [7], Lewis [8] and Lipton [9], through the introduction of generalized Fourier transform. Indeed, they demonstrated how to compute efficiently the price of derivatives contracts once the characteristic function (CF) associated with the process is known analytically.

The econophysicists' community has analyzed and developed independently models to capture the stochastic nature of volatility, with particular emphasis on the comparison of the models predictions with real market data [10–18]. The present work follows the guidelines of [11] and [16], the former focusing on the Heston's model while the latter on the Scott's model [2], also known as the exponential Ornstein-Uhlenbeck (ExpOU) model. Both these articles present the analytical characterizations of several stylized facts and in particular the tail behavior of the probability density function (PDF) of financial returns. Here we derive the closed-form formula for the CF of the linear model to which the ExpOU reduces when the level of volatility fluctuations is low. We check the goodness of the results by fast Fourier transforming the CF and comparing the numerical PDF with the Monte Carlo (MC) simulation of the stochastic models. The effectiveness in financial applications is assessed through the analysis of the German DAX30 Index.

2. The Exponential Ornstein-Ulhenbeck Model

The process we are investigating is a generalization of the stochastic volatility model originally introduced by Scott [2]. The major difference with Scott's model is the existence of the parameter ρ taking values in $[-1, 1]$. It represents the correlation between the two sources of noise in the diffusive processes and its value is crucial to obtain a non trivial behavior for the skewness of the returns PDF. The following coupled SDEs completely define the model

$$\begin{aligned} \mathrm{d}S(t) &= \mu S(t)\mathrm{d}t + m\mathrm{e}^{Y(t)} S(t)\, \mathrm{d}W_1(t)\ , \\ S(t_0) &= S_0\ , \qquad (1) \\ \mathrm{d}Y(t) &= \alpha(\gamma - Y(t))\mathrm{d}t + k\rho \mathrm{d}W_1(t) + k\sqrt{1-\rho^2}\mathrm{d}W_2(t)\ , \\ Y(t_0) &= Y_0\ , \qquad (2) \end{aligned}$$

where $\mathrm{d}W_1$ and $\mathrm{d}W_2$ are standard Brownian increments, while S_0, μ, m, Y_0, α, γ, k and ρ are constant parameters to be calibrated on real data. If we define $\sigma(t) = m\mathrm{e}^{Y(t)}$, in equation (1) we recognize the same dynamics of a Geometric Brownian motion, with a constant drift coefficient μ, while the volatility $\sigma(t)$ characterizes the amplitude of the fluctuations of S. By tuning the value of $k \geq 0$ we can strongly modify the behavior of Y. The case $k = 0$ switches off the stochastic nature of Y and it evolves deterministically towards its stationary value γ, with a characteristic time $1/\alpha$ $(\alpha > 0)$. If k is strictly greater than zero, it is known that Y follows a Gaussian process, whose mean and variance read

$$\begin{aligned} \mathbb{E}[Y] &= (Y_0 - \gamma)\mathrm{e}^{-\alpha(t-t_0)} + \gamma \overset{t\to+\infty}{\longrightarrow} \gamma\ , \\ \mathbb{E}\left[Y^2\right] - \mathbb{E}[Y]^2 &= \frac{k^2}{2\alpha}\left[\mathrm{e}^{-2\alpha(t-t_0)} + 1\right] \overset{t\to+\infty}{\longrightarrow} \beta \doteq \frac{k^2}{2\alpha}\ , \end{aligned}$$

where $\mathbb{E}[\cdot]$ denotes expectation with respect to the probability density of the Y process. By means of Itô Lemma applied to the centered logarithmic return $X(t) = \ln S(t) - \ln S_0 - \mu(t - t_0)$, equation (1) reduces to

$$\mathrm{d}X(t) = -\frac{1}{2}m^2\mathrm{e}^{2Y(t)}\mathrm{d}t + m\mathrm{e}^{Y(t)}\,\mathrm{d}W_1(t)\ . \qquad (3)$$

with boundary condition $X(t_0) = 0$.

We are interested in the characterization of $p_X(X|X_0, Y_0)$, the returns transition probability. The returns PDF $p_X(X)$ is readily obtained taking the expectation over X_0 and Y_0. In the following section we will consider the limit linear dynamics to which the ExpOU model reduces for small values of β and in the final section we will provide empirical evidences supporting the effectiveness of our approach, see also [19] for more details.

3. The Linear Model

Since β governs the level of the fluctuations of Y around its stationary value, when $\beta \ll 1$, it is possible to approximate linearly the exponential form of $\sigma(t)$; we will show how to exactly solve the associated Fokker-Planck equation. We limit our investigation to $0 < \beta \leq 0.1$, even though higher values can be explored and the effectiveness can be tested numerically via MC simulation.

When the Y process thermalizes, the random variable fluctuates around γ and we expand the exponential around this value. By defining $\bar{m} \doteq me^{\gamma}$ and introducing the random variable $Z \doteq Y - \gamma + 1$, equation (3) and equation (2) can be rewritten as

$$\begin{aligned} \mathrm{d}X &= -\frac{\bar{m}^2}{2}(2Z-1)\mathrm{d}t + \bar{m}Z\mathrm{d}W_1, \\ X(t_0) &= 0; \\ \mathrm{d}Z &= \alpha(1-Z)\mathrm{d}t + k\rho\mathrm{d}W_1(t) + k\sqrt{1-\rho^2}\mathrm{d}W_2(t), \\ Z(t_0) &= Y_0 - \gamma + 1. \end{aligned}$$

For the above model there is a non zero probability for the diffusion coefficient $\bar{m}Z$ to become negative, which can be computed and for $t - t_0 \gg 1/\alpha$ reduces to $\mathrm{Erfc}\left(\frac{1+\gamma}{\sqrt{\beta}}\right)/2$, where Erfc is the complementary error function. Practitioners usually require to keep stochastic volatility positive, however there are no financial or mathematical problems preventing it to become negative, for a discussion on this point see [20].

The Fokker-Planck backward equation satisfied by $p_X(X|X_0, Z_0)$, shortly p_X, is readily written as

$$\begin{aligned} \frac{\partial}{\partial t_0}p_X &= \frac{\bar{m}^2}{2}(2z_0-1)\frac{\partial}{\partial x_0}p_X - \alpha(1-z_0)\frac{\partial}{\partial z_0}p_X \\ &\quad -\frac{\bar{m}^2}{2}z_0^2\frac{\partial^2}{\partial x_0^2}p_X - \rho k\bar{m}z_0\frac{\partial^2}{\partial x_0 \partial z_0}p_X - \frac{k^2}{2}\frac{\partial^2}{\partial z_0^2}p_X \,. \end{aligned} \tag{4}$$

Following Heston [5], we notice that, if we assume as a final time condition the expression $e^{ix\omega}$, equation (4) is precisely the partial differential equation governing the evolution of the CF $f(\omega; X_0, Z_0)$. We try a solution of the form

$$f(\omega; X_0, Z_0) = e^{A(t-t_0,\omega)+B(t-t_0,\omega)Z_0+C(t-t_0,\omega)Z_0^2+i\omega X_0} \,. \tag{5}$$

We substitute it in equation (4) and we obtain

$$\begin{aligned} \dot{A} + \dot{B}Z_0 + \dot{C}Z_0^2 &= \tfrac{\bar{m}^2}{2}(2Z_0-1)i\omega + \tfrac{\bar{m}^2}{2}Z_0^2\omega^2 - \alpha(1-Z_0)(B+2CZ_0) \\ &\quad -\tfrac{k^2}{2}[(B+2CZ_0)^2 + 2C] - \rho k\bar{m}Z_0 i\omega(B+2CZ_0) \,, \end{aligned}$$

where the dot stays for a derivative *w.r.t.* t_0. If we collect the coefficients of the different powers of Z_0, we have the ODEs satisfied by C, B and A

$$\dot{C} = \frac{\bar{m}^2}{2}\omega^2 + 2\alpha C - 2k^2C^2 - \rho k\bar{m}i\omega 2C \,, \tag{6}$$

$$\dot{B} = \bar{m}^2 i\omega - 2\alpha C + \alpha B - 2k^2BC - \rho k\bar{m}i\omega B \,, \tag{7}$$

$$\dot{A} = -\frac{\bar{m}^2}{2}i\omega - \alpha B - \frac{k^2}{2}(B^2 + 2C) \,, \tag{8}$$

with final time conditions $C(0,\omega) = 0$, $B(0,\omega) = 0$, and $A(0,\omega) = 0$. Equation (6) is a Riccati type ODE and once it has been solved, we insert the solution in equations (7) and (8) and we integrate them out in the usual way. The explicit expression for C, B and A are quite involved. To improve the readability we introduce the auxiliary variables $b = 2\alpha(1 - i\rho\Omega)$,

$d=\sqrt{4\alpha^2\Omega^2+b^2}$, $g=(b-d)/(b+d)$, $h=i\alpha\bar{m}\Omega/k$, and $n=\alpha(b-d)/(2k^2)$, where $\Omega=k\bar{m}\omega/\alpha$. Now the three functions read

$$C(t-t_0,\omega) \;=\; \frac{b-d}{4k^2}\frac{1-\mathrm{e}^{-d(t-t_0)}}{1-g\mathrm{e}^{-d(t-t_0)}},$$

$$B(t-t_0,\omega)=2\frac{\mathrm{e}^{-\frac{1}{2}d(t-t_0)}((g+1)h-2n)+n+\mathrm{e}^{-d(t-t_0)}(n-gh)-h}{d\left(1-g\mathrm{e}^{-d(t-t_0)}\right)},$$

and

$$\begin{aligned}
A(t-t_0,\omega) \;=\; & \left[\frac{h}{2}+2\alpha\frac{n-h}{d}+2k^2\left(\frac{n-h}{d}\right)^2+\frac{b-d}{4}\right](t-t_0)\\
& -\frac{1}{2}\left[\ln\left(1-g\mathrm{e}^{-d(t-t_0)}\right)-\ln(1-g)\right]\\
& -2k^2\frac{\mathrm{e}^{-d(t-t_0)}-1}{(1-g)\left(1-g\mathrm{e}^{-d(t-t_0)}\right)}\left\{\frac{g}{d^3}\left[\frac{\alpha}{2k^2}(b+d)-h\right]^2\right.\\
& \left.+\frac{((g+1)h-2n)^2+2(n-gh)(n-h)}{d^3}+\frac{g}{d^3}(n-h)^2\right\}\\
& -4k^2\frac{(g+1)h-2n}{d^3}\left(\frac{\alpha}{k^2}b-2h\right)\frac{\left(1+g\mathrm{e}^{-\frac{d}{2}(t-t_0)}\right)\left(\mathrm{e}^{-\frac{d}{2}(t-t_0)}-1\right)}{(1-g)\left(1-g\mathrm{e}^{-d(t-t_0)}\right)}.
\end{aligned}$$

Similarly to the Heston case, non trivial problems emerge due to the multi-valued nature of complex square roots and logarithms. Due to branching effects, the CF can become discontinuous and to check the smoothness of f we follow the same arguments discussed in [21–23]. The complete characterization of f allows us to derive the analytical expression of the cumulants. Here we report the first four cumulants, under the assumptions $X_0=0$ and $Z_0=1$ (see [24] for more details about the role played by k_1-k_4 when calibrating the model)

$$\begin{aligned}
k_1 &= -\frac{m^2}{2\alpha}\tau,\\
k_2 &= \frac{m^2}{4\alpha}\left[2\frac{k^2m^2}{\alpha^2}\left(e^{-2\tau}-4e^{-\tau}-2\tau+3\right)-\frac{k^2}{\alpha}\left(e^{-2\tau}+2\tau-1\right)-4\tau\right]\\
&\quad -2\rho\frac{km^3}{\alpha^2}\left(e^{-\tau}+\tau-1\right),\\
k_3 &= 3\frac{k^2m^3}{\alpha^3}\left(e^{-2\tau}-4e^{-\tau}-2\tau+3\right)\\
&\quad +\frac{3}{2}\frac{km^3}{\alpha^2}\rho\left\{\frac{k^2m^2}{\alpha^2}\left[-e^{-2\tau}(3+2\tau)+4e^{-\tau}(3+\tau)+4\tau-9\right]+\frac{k^2}{\alpha}\left[e^{-2\tau}(1+\tau)+\tau-1\right]\right.\\
&\quad \left.+4\left[e^{-\tau}+\tau-1\right]\right\}-6\frac{k^2m^4}{\alpha^3}\rho^2\left[e^{-\tau}(2+\tau)+\tau-2\right],\\
k_4 &= 3\frac{k^2m^4}{\alpha^3}\left\{\frac{1}{2}\frac{k^2m^2}{\alpha^2}\left[-e^{-4\tau}+4e^{-3\tau}-4e^{-2\tau}(3+\tau)+4e^{-\tau}(7+2\tau)+8\tau-19\right]\right.\\
&\quad \left.+\frac{k^2}{8\alpha}\left[e^{-4\tau}+4e^{-2\tau}(1+2\tau)+4\tau-5\right]-2\left(e^{-2\tau}-4e^{-\tau}-2\tau+3\right)\right\}
\end{aligned}$$

$$-6\,\frac{k^3m^5}{\alpha^4}\rho\left[e^{-3\tau}-2e^{-2\tau}(5+2\tau)+e^{-\tau}(35+10\tau)+12\tau-26\right]$$
$$+3\,\frac{k^2m^4}{\alpha^3}\rho^2\left\{4\frac{k^2m^2}{\alpha^2}\left[-e^{-2\tau}(3+3\tau+\tau^2)+e^{-\tau}(12+6\tau+\tau^2)+3\tau-9\right]\right.$$
$$\left.+\frac{k^2}{\alpha}\left[e^{-2\tau}(3+4\tau+2\tau^2)+2\tau-3\right]-2\left[e^{-2\tau}-4e^{-\tau}(3+\tau)-6\tau+11\right]\right\}$$
$$-12\,\frac{k^3m^5}{\alpha^4}\rho^3\left[e^{-\tau}(6+4\tau+\tau^2)+2\tau-6\right],$$

with $\tau=\alpha(t-t_0)$. The first cumulant does depend only on the parameter m and this turns to be extremely useful when estimating it. If we introduce the skewness $\zeta=k_3/k_2^{3/2}$ and the kurtosis $\kappa=k_4/k_2^2$, we can study their behavior both in the short and large time limit. Indeed, when $\alpha(t-t_0)\ll 1$ we can Taylor expand their expressions finding

$$\zeta\ \sim\ 3k\rho\sqrt{t-t_0},\quad \text{and}\quad \kappa\sim 4k^2(1+2\rho^2)(t-t_0).$$

For $\alpha(t-t_0)\gg 1$, we have

$$\zeta\ \sim\ -6\frac{k^2}{\alpha^2}\frac{\left[\frac{1}{\alpha}-k\rho\left(\frac{1}{4}+\frac{m^2}{\alpha}+\frac{\alpha}{k^2}\right)+m\rho^2\right]}{\left[\left(\frac{km}{\alpha}\right)^2+\frac{1}{2}\frac{k^2}{\alpha}-1-2\rho\frac{km}{\alpha}\right]^{3/2}}\frac{1}{\sqrt{t-t_0}},$$
$$\kappa\ \sim\ 3\frac{k^2}{\alpha^2}\frac{\left[4\left(\frac{km}{\alpha}\right)^2+\frac{k^2}{2\alpha}+4-24\frac{km}{\alpha}\rho+2\rho^2\left(6\left(\frac{km}{\alpha}\right)^2+\frac{k^2}{\alpha}+6\right)-8\frac{km}{\alpha}\rho^3\right]}{\left[\left(\frac{km}{\alpha}\right)^2+\frac{1}{2}\frac{k^2}{\alpha}-1-2\rho\frac{km}{\alpha}\right]^2}\frac{1}{t-t_0}.$$

The peculiar time scaling of the previous expressions predicts no excess of skewness and kurtosis for long horizons. It would be interesting to explore the tail properties of the p_X density to verify whether the long time behavior of this model is consistent with a Gaussian dynamics.

4. Numerical Results

The Fourier anti-transform of f can not be computed analytically, but exploiting the symmetries of the CF we can evaluate it efficiently with a Fast Fourier Transform algorithm. All the results presented in the figures have been obtained with 2^{22} points on the fixed interval of integration $[0,10^3]$. To simulate the dynamics of the ExpOU and linear models, we set $\Delta t=10^{-4}$ and the number of MC paths equal to $5\cdot 10^6$. In figure 1 we report the returns histograms and the p_X curve given by the numerical Fourier anti-transform: $t-t_0=1$, $m=0.1$, $\alpha=10$, $\gamma=0$ and $Y_0=0$. Each group of three curves corresponds to β values 0.5%, 1%, 2%, 5% and 10%, from bottom to top respectively. Curves has been shifted for the sake of readability. The top frame has been obtained with a positive ρ value (0.5), which is evident from the rightward asymmetry of the curves, while, in the bottom panel, curves have large leftward asymmetry corresponding to $\rho=-0.9$. The agreement between theory prediction and MC simulation is very good for $\beta=0.5\%$, 1% and 2%, in both cases of ρ values. The returns PDFs for the exponential model are well reproduced by the ones for the linear model and, as a consequence, by the semi-analytical approximation p_X. The agreement worsen when β increases to 5% and 10%, as it can be seen by observing the fatter tail in each panel.

In figure 2 the scaling with time of mean, variance, skewness and kurtosis is shown for $\beta=0.5\%$ and 5%, $\rho=-0.9$. Then in table 1 we detail the numerical values of normalized cumulants for $\beta=2\%$ and $t-t_0=0.01,0.1,0.2,0.5,1$ (MC for the ExpOU and analytical for the linear model). Since the linear limit is an approximating dynamics of the ExpOU model, we

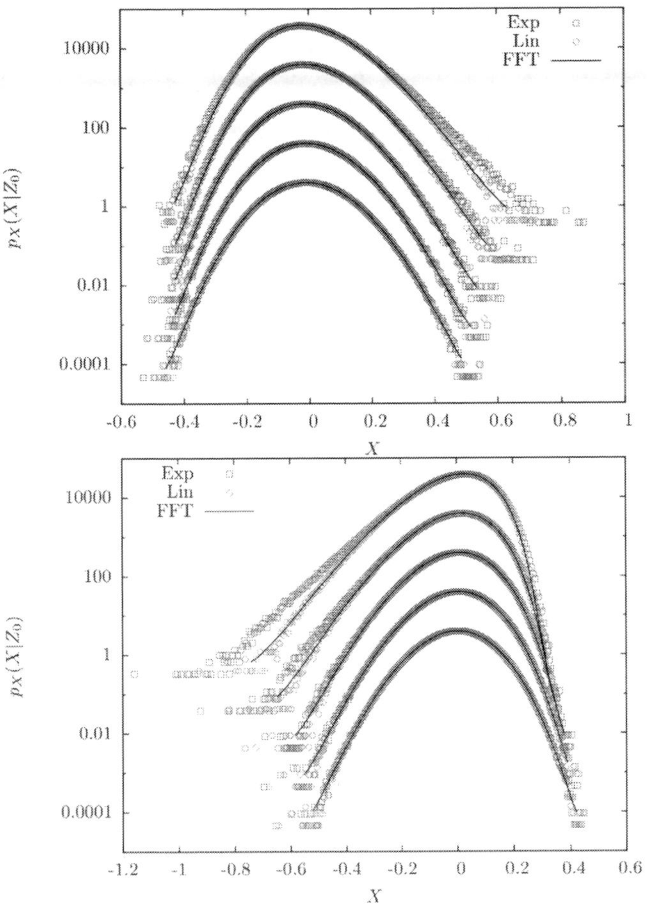

Figure 1. Returns PDF from MC simulation of the ExpOU and linear models and Fourier transform of formula (5).

do not expect necessarily a statistical compatibility between the MC estimates and the analytical values. However, we are able to quantify the relative differences among cumulants values and we can evaluate the degree of accuracy of the approximation. For example, for $\beta = 2\%$ the relative disagreement between k_2^{Exp} and k_2^{Lin} increases with time from 0.1% to 2%. As expected, this effect is enhanced by higher values of β.

5. Real data analysis

Analytical results of section 3 are tested on the German DAX30 time series, consisting of 12173 daily close prices, from 4th January 1960 until 30th June 2008. Time is measured on a yearly base, for DAX30 we have a 48.5 years time window. The model described by equations (1) and (2) depends on 8 parameters. However, S_0 corresponds to the asset spot price so the true free parameters are μ, m, Y_0, α, γ, β and ρ. In order to estimate their values we adopt the following strategy:

μ: from the discretized version of equation (1)

$$\frac{\Delta S_i}{S_i} = \mu\,\Delta t + m\,\sqrt{\Delta t}\,e^{Y_i}\epsilon_i,$$

where $\Delta S_i = S_{i+1} - S_i$ and $\epsilon_i \sim \mathcal{N}(0,1)$, we can conclude that the expectation $\langle \Delta S_i / S_i \rangle / \Delta t$ over the real data sample provides an estimate of μ; for DAX30 $\Delta t = 3.98 \times 10^{-3}$.

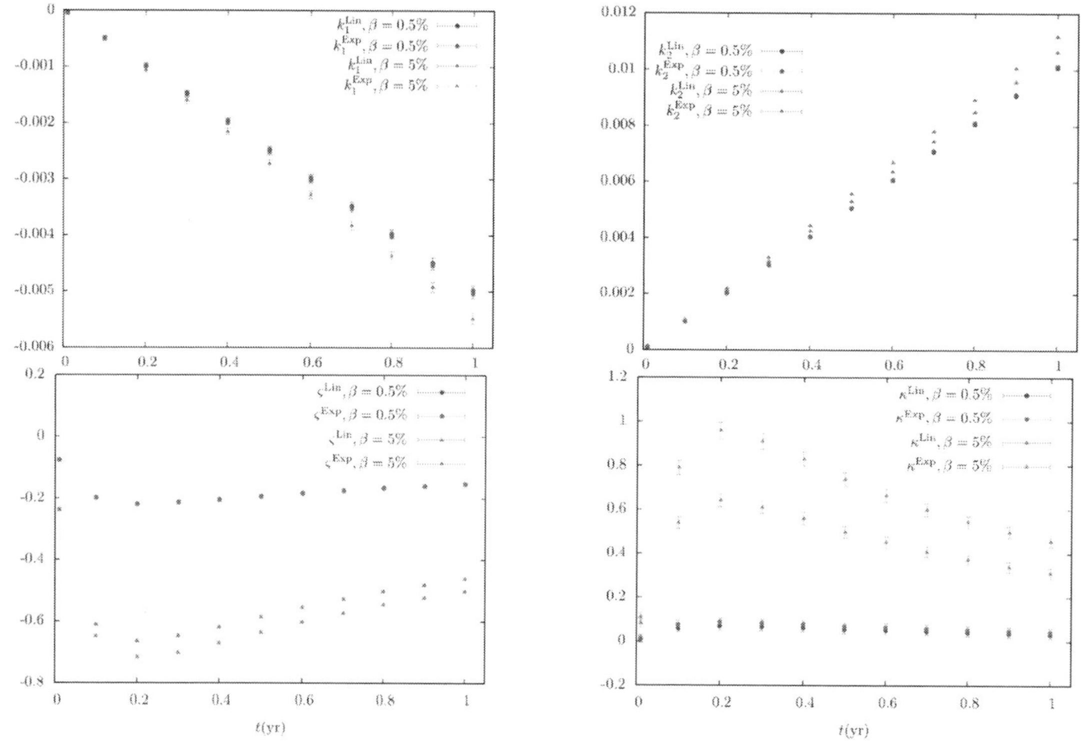

Figure 2. From top left clockwise: scaling with time of mean, variance, kurtosis and skewness.

Table 1. Scaling with time of normalized cumulants (between parenthesis the error on the last significant digit, 95% confidence level).

	$\beta = 2\%$				
$t - t_0$(yr)	0.01	0.1	0.2	0.5	1
$k_1^{\text{Exp}}(10^{-4})$	-0.50(8)	-5.0(2)	-10.1(4)	-25.8(6)	-51.8(9)
$k_1^{\text{Lin}}\ (10^{-4})$	-0.50	-5.0	-10.0	-25.0	-49.9
$k_2^{\text{Exp}}(10^{-4})$	1.004(1)	10.28(1)	20.77(2)	52.35(7)	104.8(1)
$k_2^{\text{Lin}}\ (10^{-4})$	1.003	10.16	20.43	51.36	102.9
ς^{Exp}	-0.151(4)	-0.402(4)	-0.443(4)	-0.393(4)	-0.311(4)
ς^{Lin}	-0.165	-0.393	-0.427	-0.379	-0.301
κ^{Exp}	0.04(2)	0.30(2)	0.36(2)	0.28(2)	0.17(2)
κ^{Lin}	0.04	0.22	0.26	0.21	0.13

γ, y_0: Remembering the definition $\sigma(t) = me^{Y(t)}$ and under the assumption that the volatility process has already reached the stationary state for the observed time series, it is readily proved that $\mathbb{E}\left[\sigma(t)^n\right] = (me^{\gamma})^n\, e^{\frac{n^2}{2}\beta}$. Since all the moments do not depend on Y_0 and e^{γ} is always coupled with m, we set Y_0 equal to zero and estimate $\bar{m}$.

$\bar{m}$,β: These two parameters completely specify the stationary log-normal distribution of the stochastic variable σ. To extract the distribution of the hidden variable σ from the series of financial returns, we implement the methodology described in [25].

Table 2. Estimated values for the model parameters for DAX30 Index.

	$\mu\,(\mathrm{yr}^{-1})$	y_0	γ	$\bar{m}\,(\mathrm{yr}^{-\frac{1}{2}})$	β	$\alpha\,(\mathrm{yr}^{-1})$	ρ
DAX30	7.39×10^{-2}	0	0	14.52×10^{-2}	11.16×10^{-2}	30.76	-0.54

α,ρ: Finally, to estimate α and ρ, we search for values able to reproduce the empirical scaling with time of real data skewness and kurtosis. We consider time horizons from one day to one hundred days and normalized cumulants are evaluated with standard estimators. The empirical skewness, ς_{Ph}, and kurtosis, κ_{Ph}, and corresponding errors, $\epsilon^{\varsigma}_{Ph}$ and ϵ^{κ}_{Ph} are obtained from historical returns, while ς_{MC}, κ_{MC} and associated errors, $\epsilon^{\varsigma}_{MC}$ and ϵ^{κ}_{MC} are drawn by generating 10000 MC paths. The optimal α and ρ are given by those values minimizing the sum of the normalized squared differences

$$(\alpha^*,\rho^*) \;=\; \underset{\alpha>0,\,\rho\in(-1,1)}{\operatorname{argmin}} \sum_{i=1}^{100}\left[\frac{(\varsigma_{i,Ph}-\varsigma_{i,MC})^2}{{\epsilon^{\varsigma}_{i,Ph}}^2+{\epsilon^{\varsigma}_{i,MC}}^2}+\frac{(\kappa_{i,Ph}-\kappa_{i,MC})^2}{{\epsilon^{\kappa}_{i,Ph}}^2+{\epsilon^{\kappa}_{i,MC}}^2}\right].$$

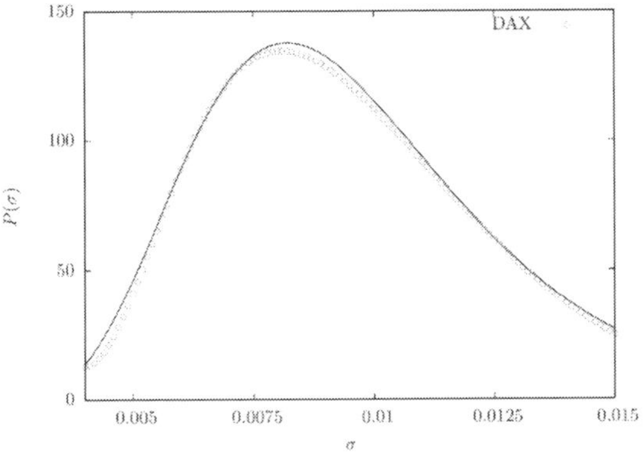

Figure 3. Volatility PDF for DAX30; the solid line corresponds to a log-normal fit.

The volatility density $P(\sigma)$ is plotted in figure 3. Tails suffer low statistics effects, but in the central region they are well fitted by a log-normal distribution

$$p(\sigma)=\frac{1}{\sqrt{2\pi}s\sigma}\exp\left[-\frac{1}{2}\left(\frac{\log\sigma-\log\sigma_0}{s}\right)^2\right].$$

The fit is performed in the range $0.0004\le\sigma\le0.015$ and gives $\log\sigma_0=-4.492\pm0.001$ and $s=0.334\pm0.001$. Table 2 details the corresponding values for $\bar{m}=\sigma_0/\sqrt{\Delta t}$ and $\beta=s^2$. Finally, we present the probability densities for DAX30, see figure 4, computed according to the models we have discussed in this work. The first row is obtained by setting the time horizon equal to 25 trading days, while the second one corresponds to the 45 trading days horizon. The densities are presented both in log-linear and linear scales, in order to allow the reader to appreciate the behavior on the tails and in the central region, respectively. The boxed-lines represent the empirical histograms, while black points correspond to the histograms from the MC simulation of the ExpOU model. Moreover, we report the Fourier transform of the CF for the linear model (solid black line) and the Normal Maximum Likelihood fit (solid red line). The Normal approximation is scarcely representative of the true empirical density, while the exponential model captures in a quite effective way the leftward asymmetry, the fatter tails and the narrower central region. Figure 4 clearly shows the good performance of the linear model for the DAX30 Index.

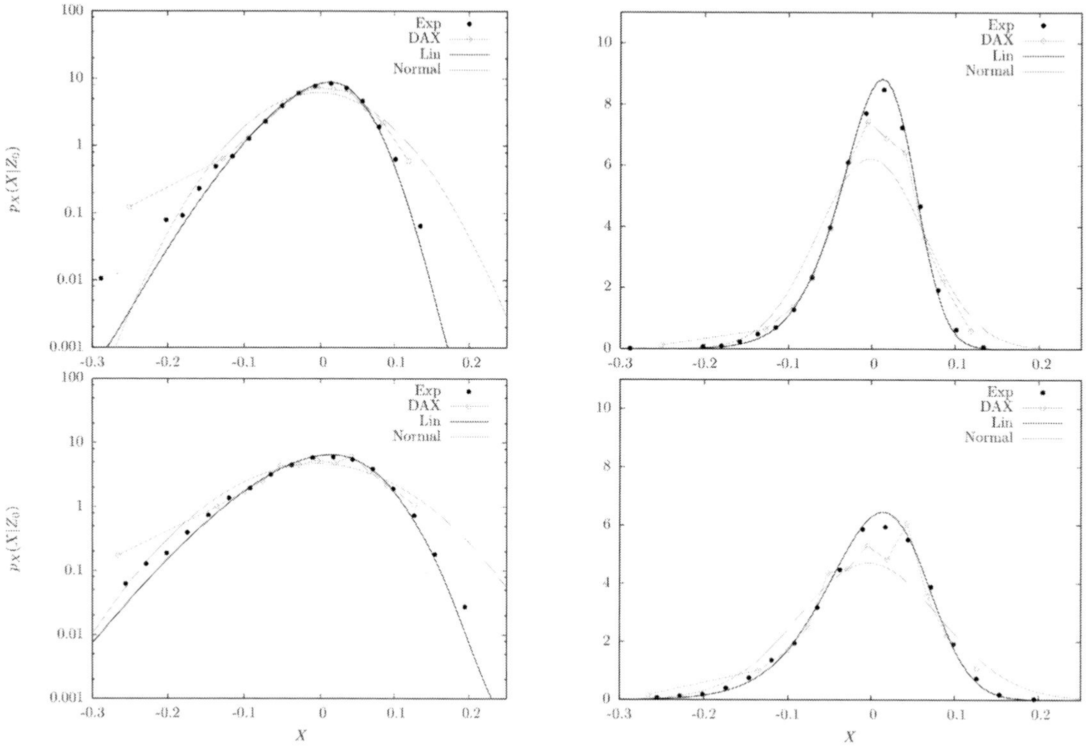

Figure 4. DAX30 index returns distributions for two different time horizons, 25 trading days (first row) and 45 trading days (second row).

6. Conclusions

The ExpOU dynamics has been widely studied in the econophysics literature from the point of view of multi-time scale properties, leverage effect and stationary volatility distribution [11–13, 16, 17, 26]. However, since our interest is driven by the possibility of a financial application in the context of option pricing and risk management we focused on the analytical characterization of the CF. Several attempts to obtain an approximated closed-form expression for the PDF can be found in [16, 19, 27]. In this work we explored the possibility to approximate linearly the exponential form of the volatility under a low fluctuations regime of the secondary process Y. Following Heston [5], we solved exactly the Fokker-Planck backward equation associated with the linear model and in Section 3 we reported the full expression of the CF. The numerical PDF obtained by anti-transforming our analytical CF was checked against the MC simulation of the linear process, finding a perfect agreement. The analytical predictions were tested over a data set of market returns. We detailed the performances for the German DAX30 Index, confirming the capability of the linear model to capture the statistical properties of the returns distribution for moderate values of β and over appropriate time horizons. The results discussed in this paper have been exploited successfully for the characterization of the ExpOU model and its linear limit in a risk-neutral framework, see [24].

Acknowledgments

G. B. wishes to thank THIC & APFA7 Organizing Committee for providing financial support to partially cover travelling expenses.

References

[1] Bachelier L 1900 *Ann. Sci. École Norm. Sup.* **17** 21

[2] Scott L 1987 *J. Finan. Quant. Anal.* **22** 419

[3] Hull J and White A 1987 *J. Finance* **42** 281

[4] Stein E and Stein J 1991 *Rev. Finan. Stud.* **4** 727

[5] Heston S L 1993 *Rev. Finan. Stud.* **6** 327

[6] Ball C and Roma A 1994 *J. Finan. Quant. Anal.* **29** 589

[7] Carr P and Madan D 1999 *J. Comput. Finance* **2** 61

[8] Lewis A L 2001 A simple option formula for general jump-diffusion and other exponential Lévy processes (*Preprint* `http://ssrn.com/abstract=282110`)

[9] Lipton A 2001 *Mathematical Methods For Foreign Exchange: A Financial Engineer's Approach* (Singapore: World Scientific Publishing)

[10] Bacry E, Delour J and Muzy J F 2001 *Phys. Rev. E* **64** 026103

[11] Dragŭlescu A A and Yakovenko V M 2002 *Quant. Finance* **2** 443

[12] Masoliver J and Perelló J 2002 *Int. J. Theoretical Appl. Finance* **5** 541

[13] Masoliver J, Perelló J and Anento N 2004 *Physica A* **344** 134

[14] Borland L, Bouchaud J P, Muzy J F and Zumbach G 2005 *Wilmott*

[15] Borland L and Bouchaud J P 2005 On a multi-timescale statistical feedback model for volatility fluctuations (*Preprint* `arXiv:physics/0507073v1 [physics.soc-ph]`)

[16] Masoliver J and Perellò J 2006 *Quant. Finance* **6** 423

[17] Cisana E, Fermi L, Montagna G and Nicrosini O 2007 A comparative study of stochastic volatility models (*Preprint* `arXiv:0709.0810v1 [q-fin.ST]`)

[18] Bonanno G, Valenti D and Spagnolo B 2007 *Phys. Rev. E* **75** 016106

[19] Bormetti G, Cazzola V, Montagna G and Nicrosini O 2008 *J. Stat. Mech.* JSTAT11(2008)013

[20] Lipton A and Sepp A 2008 *J. Physics A* **41** 344012

[21] Albrecher H, Mayer P, Schoutens W and Tistaert J 2007 *Wilmott* 83

[22] Kahl C and Jackel P 2005 *Wilmott* 94

[23] Lord R and Kahl C 2006 Why the rotation count algorithm works (Tinbergen Institute Discussion Paper)

[24] Bormetti G, Cazzola V and Delpini D 2009 Option pricing under Ornstein-Uhlenbeck stochastic volatility (*Preprint* `arXiv:0905.1882 [q-fin.PR]`)

[25] Pasquini M and Serva M 2000 *Eur. Phys. J. B* **16** 195

[26] Bouchaud J P and Potters M 2000 *Theory of Financial Risk and Derivative Pricing: From Statistical Physics to Risk Management* (Cambridge: Cambrigde University Press)

[27] Perellò J, Sircar K R and Masoliver J 2008 *J. Stat. Mech.* JSTAT06(2008)010

Dependence of the Number of Dealers in a Stochastic Dealer Model

Kenta Yamada

Department of Computational Intelligence and Systems Science, Interdisciplinary Graduate School of Science and Engineering, Tokyo Institute of Technology, 4259 Nagatsuta-cho, Midori-ku, Yokohama 226-8502, Japan

E-mail: yamada@smp.dis.titech.ac.jp

Hideki Takayasu

Sony Computer Science Laboratories, 3-14-13 Higashi-Gotanda, Shinagawa-ku, Tokyo 141-0022, Japan

Misako Takayasu

Department of Computational Intelligence and Systems Science, Interdisciplinary Graduate School of Science and Engineering, Tokyo Institute of Technology, 4259 Nagatsuta-cho, Midori-ku, Yokohama 226-8502, Japan

Abstract. We numerically analyze an artificial market model consisted of N dealers with time dependent stochastic strategy. Observing the change of market price statistics for different values of N, it is shown that the statistical properties are almost same when the dealer number is larger than about 30.

1. Introduction

Through intensive analysis of high frequency data, stylized facts have been established about fluctuation of financial markets especially for the foreign exchange markets which are the largest free markets in the world and open 24 hours [1]- [7].

♯ **1** The power law distributions of volatility, or the absolute value of price changes.

♯ **2** The autocorrelation function of price changes which takes a negative value at the first tick and otherwise it is close to 0.

♯ **3** Long time correlation of volatility.

♯ **4** Diffusion properties: Abnormal diffusion in short time scale and normal diffusion in the large time scale.

♯ **5** Non-Poissonian transaction intervals characterized by a non-exponential distribution.

From ♯1 to ♯4 in the above list, these properties are derived by analyzing time series of market prices while ♯5 comes from analysis of transaction intervals. There are many market models which satisfy some of the above empirical laws[8]-[23].

For the purpose of clarifying the relation between the above statistical properties and dealers' action in the market, we already introduced deterministic dealer models and a stochastic dealer model [24]-[27]. In these series of study of market model construction, we started with the most basic model which produces non-correlated fluctuations both for price changes and transaction intervals. This model, named as "model 1", can be recognized as a random number generator in the market, however, this simple model does not satisfy distribution of transaction intervals and volatility.

In order to realize the fat-tail property of transaction intervals (♯5), it is important to introduce a kind of modulation of dealers' activity called, "self modulation", to be proportional to the market's activity which is defined by the number density of transactions. The model with this revision is named as "model 2".

As for power law distribution of price changes (♯1), we need to add a term describing the effect of following trends, which are defined by mean gradients of latest price changes, and we call the revised model as "model 3". In the model-4, we combine model 2 and model 3 and we show model 4 fulfills all major statistical properties.

The goal of this paper is to observe the effect of number of dealers in these models and to reconstruct market stylized facts. In the next section we construct our market model following the style of former model construction, namely, we start with "model 1", then introduce "model 2" and "model 3", and finally combining all terms in the subsection "model 4".

2. Construction of the dealer models

2.1. Model 1

We construct an artificial market consisted of N dealers modeling foreign exchange or stock markets. Every dealer shows his limit prices of bid and ask. A bid price is the price to buy while an ask price is the price to sell. It is natural that the ask price is always higher than a bid price for each dealer because each dealer wants to get some margin by transactions. In the model, we assume that the spread, which is defined by the ask price minus the bid price, to be a constant, L, for all the dealers for simplicity. We define the ith dealer's price at time t, $p_i(t)$, by the mean value between the bid and ask prices, namely, the bid and ask prices are given by $p_i(t) - L/2$ and $p_i(t) + L/2$, respectively. When the ith dealer's bid price, $p_i(t) - L/2$, is larger or equal to the jth dealer's ask price, $p_j(t) + L/2$, then a transaction occurs between these dealers. This condition is represented by an inequality, $|p_i(t) - p_j(t)| \geq L$. The market's deal price $P(t)$ is then given by $\{p_i(t) + p_j(t)\}/2$ and after the transaction we assume that both dealers' prices are set to be equal to this market's deal price.

The time evolution of each dealers' price is given by the following rule:

$$\begin{aligned} p_i(t + \Delta t) &= p_i(t) + c f_i(t) \qquad (1) \\ f_i(t) &= \begin{cases} +\Delta p & (\text{prob. } 1/2) \\ -\Delta p & (\text{prob. } 1/2) \end{cases} \qquad i = 1, 2, \cdots, N, \end{aligned}$$

where Δt is the time step, c is a constant representing the step size of random motion of each dealer, $f_i(t)$ takes either Δp or $-\Delta p$ randomly with equal probability, and Δp is a positive constant.

We show market price fluctuations and transaction intervals for some values of N in Fig.1 simulated with the following parameters; $L = 0.1$, $c = 0.01$, $\Delta p = 0.01$ and $\Delta t = (\Delta p)^2$. For larger number of dealers, fluctuations of both transaction intervals and price changes become smaller as clearly seen from these figures.

In Fig.2, we show the averages of transaction intervals and volatility as a function of N keeping other parameters constants. An average value of transaction intervals is proportional to N^{-1} and the volatility is approximately proportional to $N^{-0.5}$. The figures of Fig.3 are statistical properties of model 1 and the number from (i) to (v) in figure is correspond to the number of statistical property in introduction. In these simulation we use following parameters; $c = 0.01$, $\Delta p = 0.01$ and $\Delta t = (\Delta p)^2$. When N equals 2, $L = 0.015$ while the value of L for $N \neq 2$ is determind by the relation $\sqrt{N}/L \sim 66.7$, which is derived from two results; one is the relation between transaction intervals and number of dealers as seen in Fig.2, and the other is that average of transaction interval is proportional to L^2. Note that we use same parameters in model 2, 3 and 4.

From the results of Fig.3 (i) and (v), the distribution of volatility and transaction intervals does not have long correlation unlike real market. And autocorrelation function of volatility also does not have long correlation(Fig. 3 (iii)). However the autocorrelation function of price changes takes a non-trivial negative value at one tick in the case of $N \geq 3$, and its value converges for dealer number larger than 32 as shown in Fig.3 (ii). And this negative correlation cause abnormal diffusion in short time scale as shown in Fig.3 (iv). In the following section we compare these statistical properties of each model by using same form.

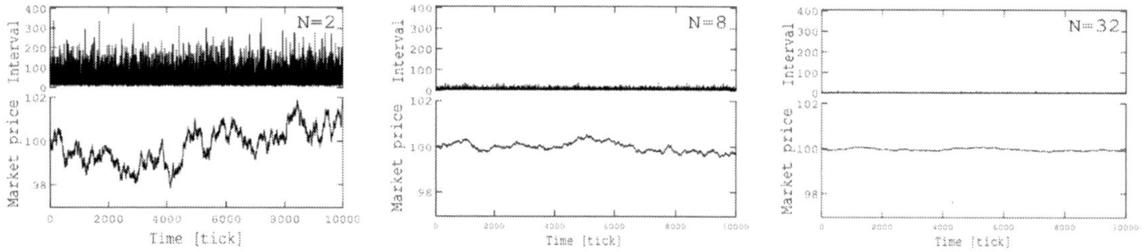

Figure 1. Simulation results of model 1 for three cases of N in the same axis scale. The upper figures are transaction intervals and lower figures are market prices.

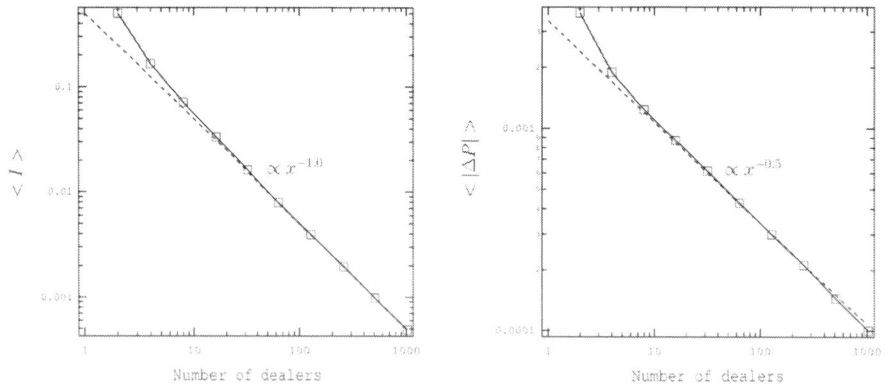

Figure 2. Average values of transaction intervals and volatility as functions of N.

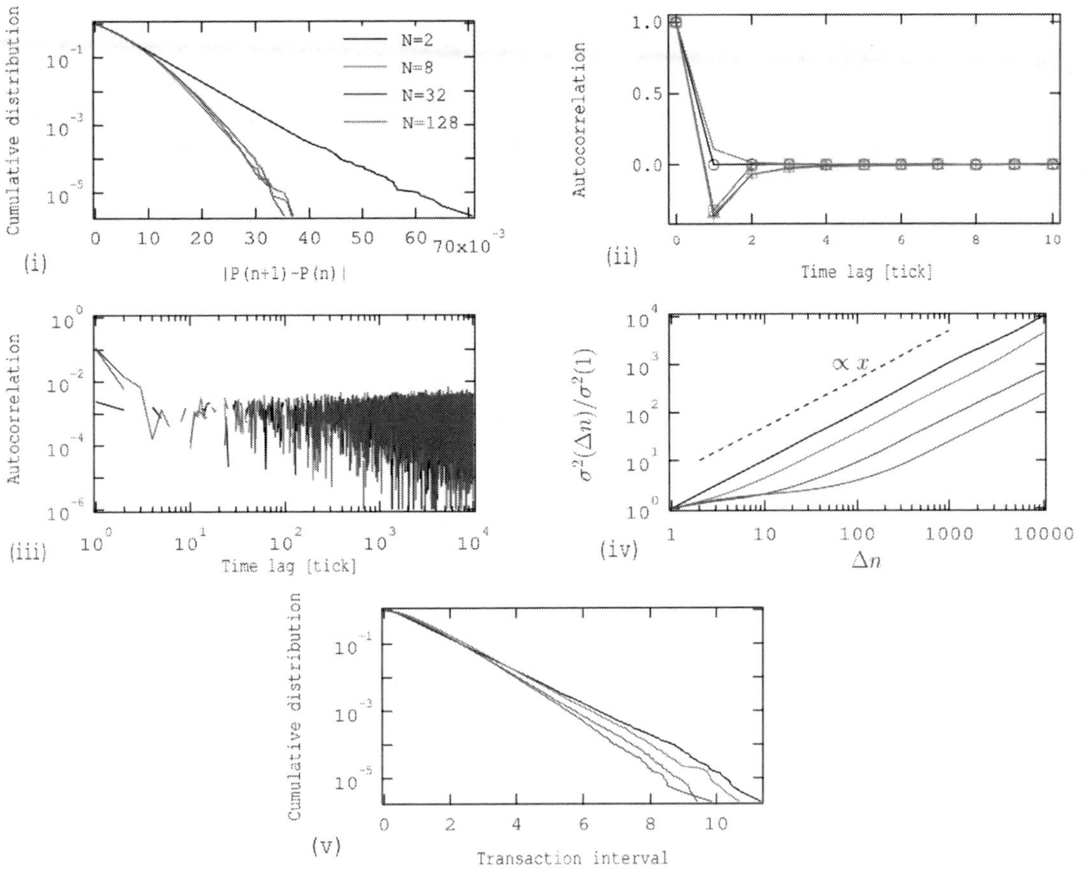

Figure 3. Statistical properties of model 1. (i): Cumulative distribution of volatility in semi-log plot. (ii): Autocorrelation function of price changes. (iii): Autocorrelation function of volatility in log-log plot. (iv): Diffusion properties. (v): Cumulative distribution of transaction intervals.

2.2. model 2

In this subsection we focus on transaction intervals. It is known that transaction intervals in real markets do not follow simple Poisson process and its statistics is characterized by a fat-tailed distribution [7]. Here, we generalize the coefficient c in Eq.(1) to be time dependent as follows:

$$p_i(t+\Delta t) = p_i(t) + c(n) f_i(t) \quad i = 1, 2, \cdots, N, \tag{2}$$
$$f_i(t) = \begin{cases} +\Delta p & \text{(prob. } 1/2) \\ -\Delta p & \text{(prob. } 1/2) \end{cases} .$$

where n denotes the tick number which counts the number of deals, and

$$c(n) = c_1 \sqrt{\frac{< I >_{c=c_1}}{< I >_\tau}}. \tag{3}$$

where c_1 is a positive constant, $< I >_{c=c_1}$ is the mean value of transaction intervals of model 1 in the case when the parameter c equals to a constant c_1, and $< I >_\tau$ is the moving average of transaction intervals averaged over the latest τ seconds defined as $< I >_\tau = \frac{1}{K}\sum_{k=0}^{K-1} I(n-k)$, with K being the number of deals within the time interval $[t-\tau, t]$, and $I(n)$ denotes the nth transaction interval, and in the case that $I(n) > \tau$, we set $< I >_\tau = I(n)$. It is known from Eq.(3) that when $< I >_{c=c_1}$ over $< I >_\tau$ is large, the value of $c(n)$ is large, and the dealers tend to make large changes in their prices. As a result there is a tendency that occurrence of transactions becomes higher when there were more transactions in the latest time interval $[t-\tau, t]$ which belongs to a kind of self-modulation effect[28]. By this effect the distribution of transaction intervals in the model 2 is modified to have fat-tails as shown in Fig.4 (v) for various N. While the statistical properties of price as shown in from (i) to (v) is unchanged from the model 1. In this simulation, we set parameters as following; $c_1 = 0.01$, $\Delta p = 0.01$, $\Delta t = (\Delta p)^2$ and $\tau = 150$. And we set L=0.015 when $N = 2$ and $L = \sqrt{N}/66.7$ when $N \neq 2$.

2.3. model 3

In this section we add so-called the trend-follow effect to the model 1. We modify the dynamics as follows;

$$p_i(t+\Delta t) = p_i(t) + d_i(n) < \Delta P >_M \Delta t + c f_i(t), \tag{4}$$
$$f_i(t) = \begin{cases} +\Delta p & \text{(prob. } 1/2) \\ -\Delta p & \text{(prob. } 1/2) \end{cases} \quad i = 1, 2, \cdots, N.$$

where the parameter $d_i(n)$ characterizes the ith dealer's strategy to be explained in more detail. $< \Delta P >_M$ is the moving average with a linearly decaying weight which is defined by the following equation,

$$< \Delta P >_M = \frac{2}{M(M+1)} \sum_{k=0}^{M-1} (M-k)\Delta P(n-k). \tag{5}$$

where $\Delta P(n) = P(n) - P(n-1)$ is the price change between the $n-1$th tick and the nth tick.

A dealer with a positive d_i is called the trend-follower who predicts market prices will go up (or down) when the moving average value Eq.(5) is positive (or negative). On the other hand a dealer with a negative d_i is called the contrarian who forecasts that the market prices will turn and go down (or up) when the moving average value is positive (or negative). The strength

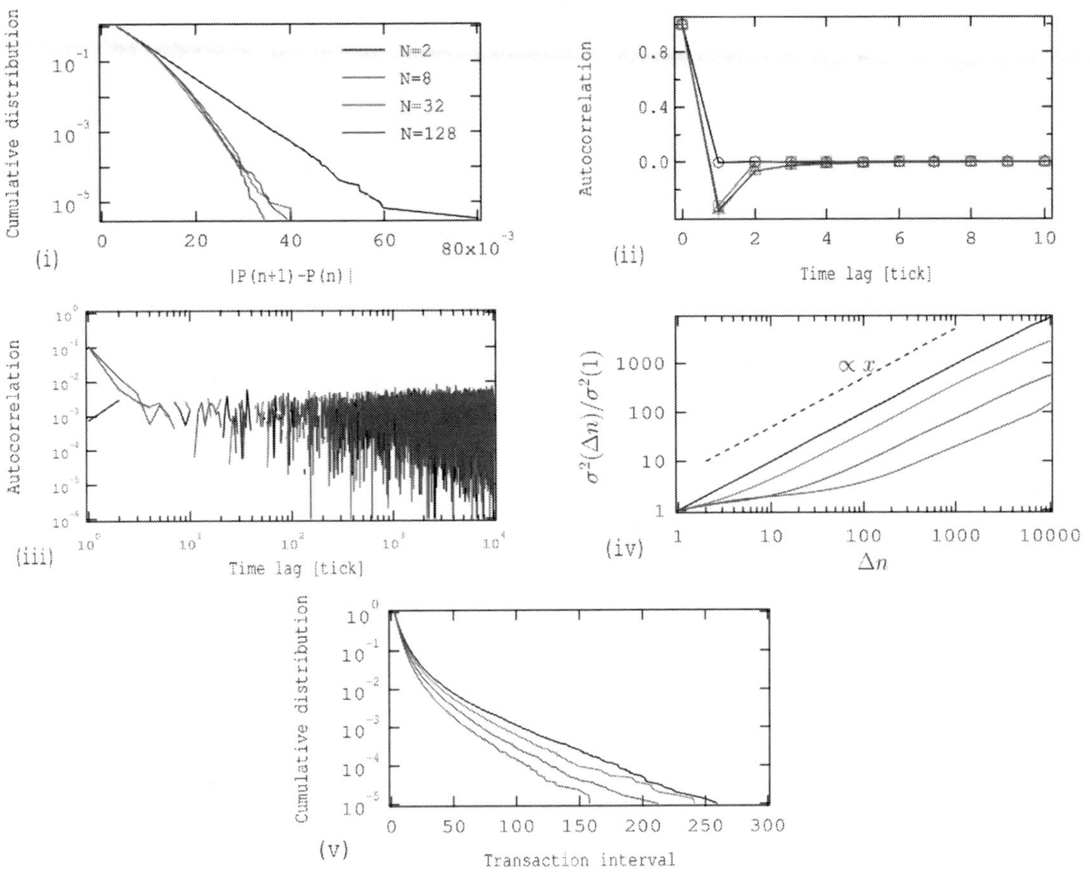

Figure 4. Statistical properties of model 2. (i): Cumulative distribution of volatility in semi-log plot. (ii): Autocorrelation function of price changes. (iii): Autocorrelation function of volatility in log-log plot. (iv): Diffusion properties. (v): Cumulative distribution of transaction intervals.

of such trend dependence is characterized by the absolute value of d_i. We assume that each dealers' strategy is changing at each market deal by the following stochastic process;

$$d_i(n) = \overline{d}(n) + \Delta d_i(n), \tag{6}$$

$$\begin{aligned} \overline{d}(n+1) &= (1-e_0)\cdot\overline{d}(n) + \phi(n), \\ \phi(n) &= \begin{cases} +0.01 & \text{(prob. } 1/2) \\ -0.01 & \text{(prob. } 1/2) \end{cases}, \end{aligned} \tag{7}$$

where $\overline{d}$(n) gives the mean strategy of traders at the nth tick, e_0 is a positive constant less than 1, Δd_i are random numbers following a normal distribution with 0 mean and the standard deviation 0.1.

In our former study we proved in the special case of $N = 2$ with a constant $d_i = d$ that the dealer model with the trend-follow effect produces time series of prices which are characterized by a market potential force, and the quadratic potential's coefficient b in the theory of PUCK analysis is linearly related to the parameter of the dealers' strategy $d_i(n)$ in Eq.(4)[27];

$$< b >= -d\left(\frac{L}{c}\right)^2. \tag{8}$$

This result implies that when dealers are trend follower ($d > 0$), the market is unstable ($b < 0$) and when dealers are contrarians ($d < 0$), the market is stable ($b > 0$). As Eq.(7) is an Auto Regressive process around $\overline{d}(n) = 0$, the corresponding market potential function also follows an Auto Regressive process around $b = 0$[29].

Fig.5 shows simulation results in the case of $N = 2$. Here, the model's parameters are given as follows; $L = 0.015, M = 1, \Delta p = 0.01, \Delta t = (\Delta p)^2, e_0 = 8.0 \times 10^{-4}$. Fig.5(a) shows the fluctuation of $\overline{d}$. When $\overline{d}$ is positive the market is dominated by the trend followers while when $\overline{d}$ is negative the market is dominated by contrarians. Fig.5(b) shows the corresponding market prices and (c) represents the price changes of market prices.

Fig.6 shows statistical properties of simulation results for different number of dealers by the model 3, we can find power law distribution of price changes (Fig.6 (i)), a quick decay of autocorrelation function of price changes (Fig.6 (ii)), long time correlation of volatility (Fig.6 (iii)), and abnormal diffusion in the short time scale and normal diffusion corresponding to simple random walk in the large time scale (Fig.6 (iv)). However, distribution of transaction intervals in Fig.6 (v) is the same as model 1 because we do not add self-modulation effect to the model 3.

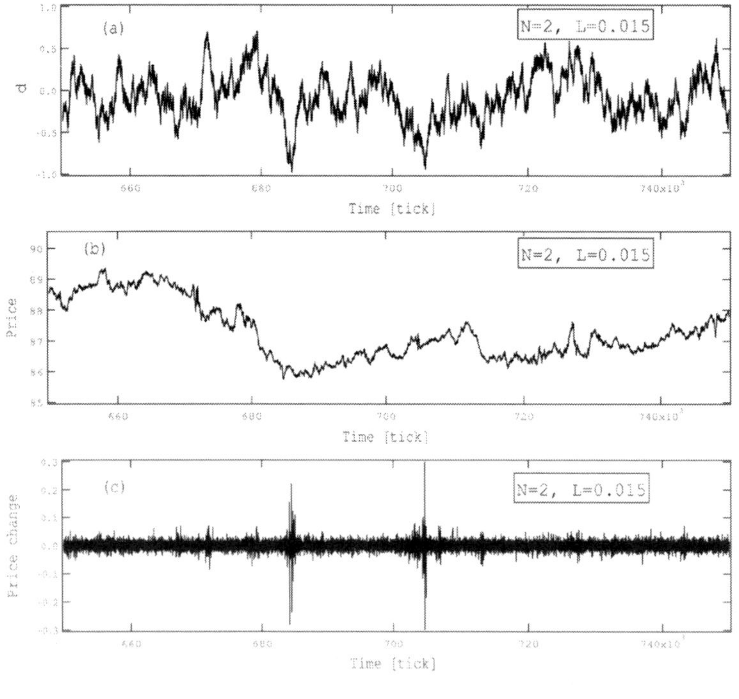

Figure 5. Simulation results in the case of $N = 2$. (a): Fluctuation of $\overline{d}$. (b): An example of market price evolution. (c): Corresponding price changes.

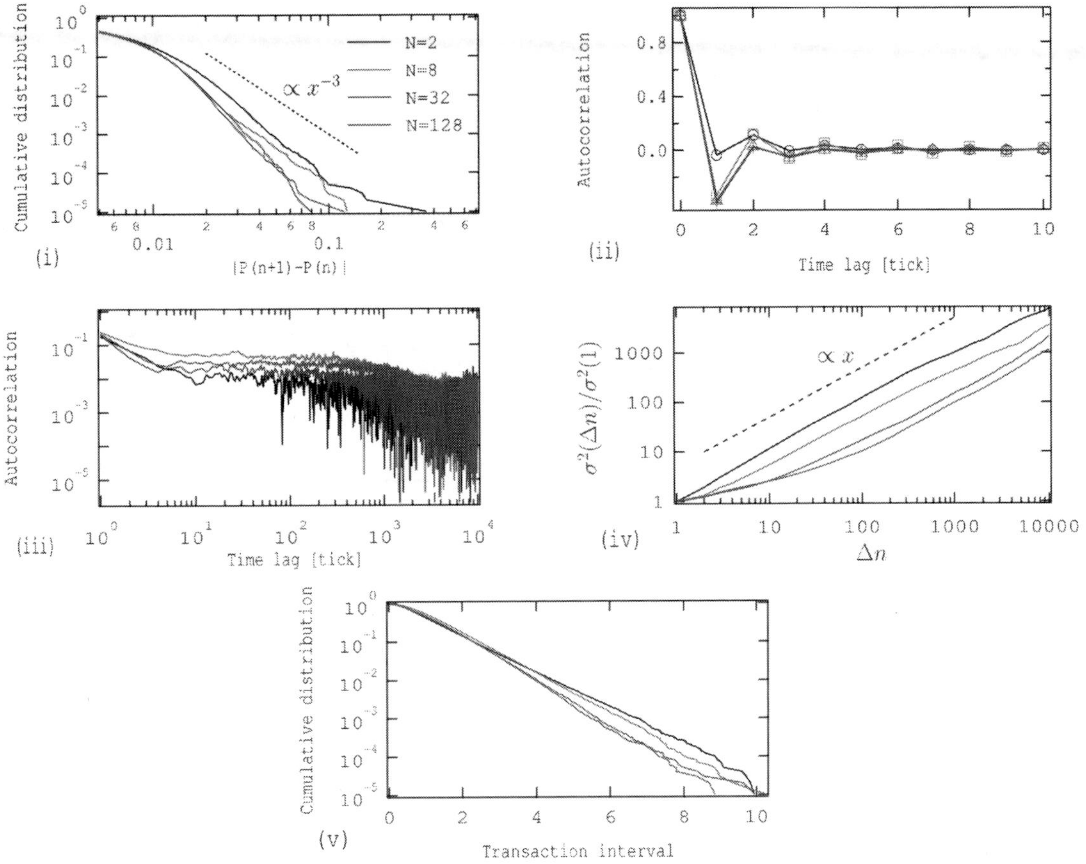

Figure 6. Statistical properties of model 3. (i): Cumulative distribution of volatility in log-log plot. (ii): Autocorrelation function of price changes. (iii): Autocorrelation function of volatility in log-log plot. (iv): Diffusion properties. (v): Cumulative distribution of transaction intervals.

2.4. model 4

Here, we combine the time modulation effect of model 2 and the trend-follow effect of model 3. The time evolution equation is given as follows;

$$p_i(t+\Delta t) = p_i(t) + d_i(n) < \Delta P >_M \{c(n)/c_1\}^2 \Delta t + c(n) f_i(t), \tag{9}$$

$$f_i(t) = \begin{cases} +\Delta p & (\text{prob. } 1/2) \\ -\Delta p & (\text{prob. } 1/2) \end{cases} \quad i = 1, 2, \cdots, N.$$

$d_i(n)$, $< \Delta P >_M$ and $c(n)$ are defined by Eq.(6), Eq.(5) and Eq.(3) respectively. And we set parameters as following; $c_1 = 0.01$, $\Delta p = 0.01$ and $\Delta t = (\Delta p)^2$, $M = 1$, $e_0 = 8.0 \times 10^{-4}$ and $\tau = 150$. And we set $L = 0.015$ when $N = 2$ and $L = \sqrt{N}/66.7$ when $N \neq 2$. By this time evolution rule all empirical findings, ♯1 − ♯5, are fulfilled as seen in Fig.7.

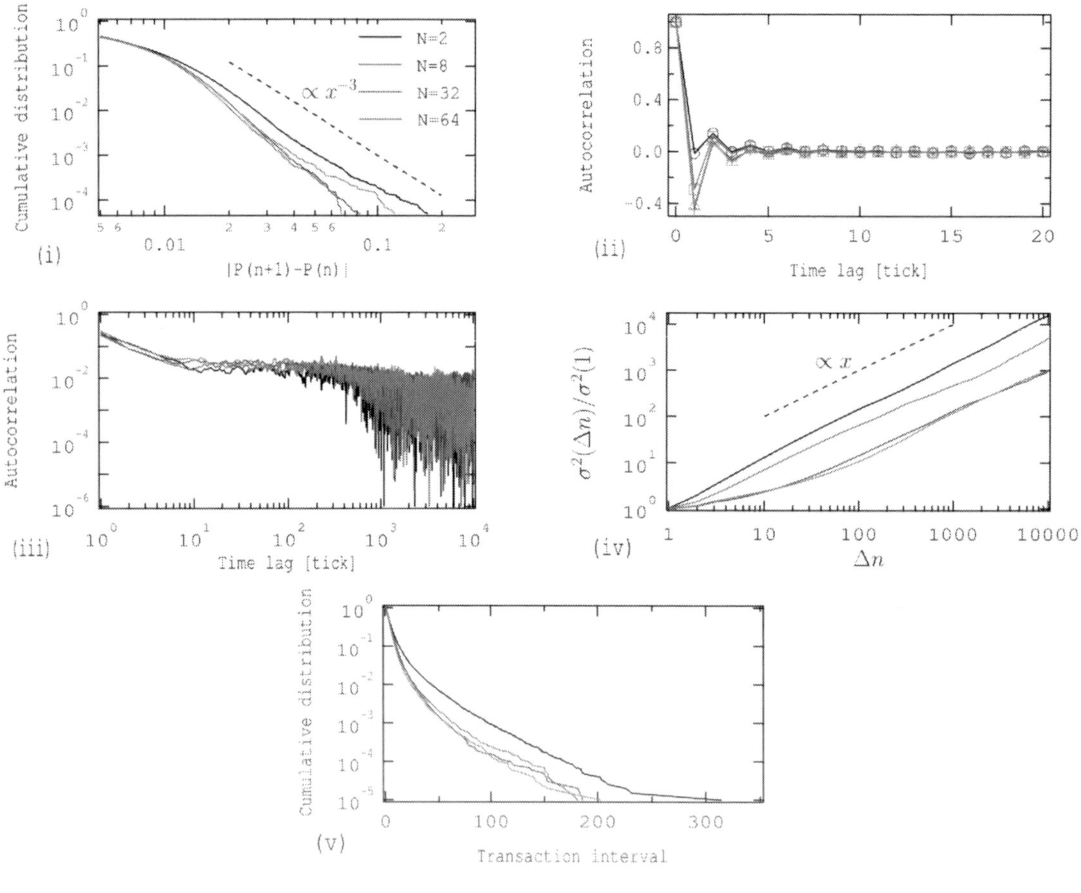

Figure 7. Statistical properties of model 4. (i): Cumulative distribution of volatility in log-log plot. (ii): Autocorrelation function of price changes. (iii): Autocorrelation function of volatility in log-log plot. (iv): Diffusion properties. (v): Cumulative distribution of transaction intervals.

3. Concluding Discussion

We studied 4 types of dealer models with various numbers of dealers. In the model 1 we confirm that the fluctuations of transaction intervals decreases for larger number of dealers proportional to N^{-1}, and the volatility shrinks proportional to $N^{-0.5}$, respectively. We found that there is a non-trivial autocorrelation in price changes at short time when N is equal to or greater than 3. In the model 2, we reproduce transaction distribution with a fat-tail by adding feedback effect of past transaction intervals for any number of dealers. In the model 3, we added dealers' trend follow effect to model 1 and reproduced other statistical properties, the power law distribution of price changes, the long time autocorrelation of volatility and the abnormal diffusion property. By combining the model 2 and model 3 to make the model 4, all empirical statistical properties are fulfilled automatically.

In the real foreign exchange markets almost all dealers belong to large international banks and the total number of active dealers is expected to be not so large, about the order of hundreds even in the case of Dollar-Euro markets which is the largest financial market in the world. The number of active dealers in other markets may be smaller, however, if the number is of order of tens we may expect that the basic statistical properties will almost be the same as those of larger markets.

References

[1] R. N. Mantegna and H. E. Stanley, An Introduction to Econophysics-Correlation and Complexity in Finance-, Cambridge University Press, Cambridge (2000).

[2] J.-P. Bouchaud and M. Potters, Theory of Financial Risks (Cambridge University Press, Cambridge, 2000).

[3] T. Mizuno, S. Kurihara, M. Takayasu and H. Takayasu, Physica A 324, 296-302 (2003).

[4] P. Gopikrishnan, M. Meyer, L. A. N. Amaral, V. Plerou and H. E. Stanley, Eur. Phys. J. B 3, 139-140, (1998).

[5] Y. Liu, P. Gopikrishnan, P. Cizeau, M. Meyer, C.-K. Peng, and H. E. Stanley, Phys. Rev. E 60, 1390-1400 (1999).

[6] T. Ohnishi, H. Takayasu, T. Ito, Y. Hashimoto, T. Watanabe, M. Takayasu" Journal of Economic Interaction and Coordination, Vol. 3, Number 1, 99-106 (2008).

[7] M. Takayasu, H. Takayasu and M.P. Okazaki, Proceedings of "Empirical Science of Financial Fluctuations" in Tokyo. (edited by H. Takayasu, Springer, 18-26, 2001).

[8] S. Maslov, Physica A 278, 571-578 (2000).

[9] F. Slanina, Phys. Rev. E 64, 056136 (2001).

[10] M. G. Daniels, J. D. Farmer, L. Gillemot, G. Iori, and E. Smith, Phys. Rev. Lett. 90, 108102 (2003) E. Smith, J. D. Farmer, L. Gillemot, and S. Krishnamurthy, Quant. Finance 3, 481 (2003).

[11] A. Svorenčík and F. Slanina, Eur. Phys. J. B 57, 453-462 (2007).

[12] J. Maskawa, Physica A 382, 172-178 (2007).

[13] S. Mike and J.D. Farmer, Journal of Economic Dynamics and Control, 32, 200-234 (2008).

[14] T. Lux and M. Marchesi, Nature 397, 498-500 (1999).

[15] V. Alfi, M. Cristelli, L.Pietronero and A. Zaccaria, Eur. Phys. J. B 67, 385 (2009).

[16] V. Alfi, M. Cristelli, L.Pietronero and A. Zaccaria, Eur. Phys. J. B 67, 399 (2009).

[17] D. Challet, M. Marsili, and R. Zecchina, Phys. Rev. Lett. 84,1824 (2000).

[18] D. Challet and M. Marsili, Phys. Rev. E 68, 036132 (2003).

[19] A. De Martino, I. Giardina, A. Tedeschi, and M. Marsili, Phys. Rev. E 70, 025104(R) (2004).

[20] F. Ren, B. Zheng, T. Qiu, S. Trimper, Phys. Rev. E 74, 041111 (2006).

[21] Krawiecki, A., Hołyst, J.A. and Helbing, D., Phys. Rev. Lett. 89, 158701(2002).

[22] T. Kaizoji, Physica A: Statistical Mechanics and its Applications 370, 109-113(2006).

[23] W. Zhou and D. Sornette, Eur. Phys. J. B 55, 175-181 (2007).

[24] H.Takayasu, H. Miura, H. Hirabayashi and K. Hamada, Physica A 184, 127-134 (1992).

[25] A. Sato and H.Takayasu, Physica A 250, 231-252 (1998).

[26] K. Yamada, H. Takayasu and M. Takayasu, Physica A 382, 340-346 (2007).

[27] K. Yamada, H. Takayasu and M. Takayasu, Phys. Rev. E 79, 051120 (2009).

[28] M. Takayasu, H. Takayasu, Physica A324, 101-107 (2003).

[29] M. Takayasu, K. Watanabe, T. Mizuno and H. Takayasu, in preparation.

Time scale defined by the fractal structure of the price fluctuations in foreign exchange markets

Yoshiaki Kumagai

Faculty of Education and Integrated Arts and Sciences, Waseda University 1-6-1, Nishi-Waseda, Shinjuku-ku, Tokyo 169-8050, Japan

E-mail: `kumagai-y@waseda.jp`

Abstract. In this contribution, a new time scale named C-fluctuation time is defined by price fluctuations observed at a given resolution. The intraday fractal structures and the relations of the three time scales: real time (physical time), tick time and C-fluctuation time, in foreign exchange markets are analyzed. The data set used is trading prices of foreign exchange rates; US dollar (USD)/Japanese yen (JPY), USD/Euro (EUR), and EUR/JPY. The accuracy of the data is one minute and data within a minute are recorded in order of transaction. The series of instantaneous velocity of C-fluctuation time flowing are exponentially distributed for small C when they are measured by real time and for tiny C when they are measured by tick time. When the market is volatile, for larger C, the series of instantaneous velocity are exponentially distributed.

1. Introduction

There are at least four candidates for the proper time scale to use for analyzing market prices; physical time (real time), trading time , the number of trade, and cumulated trading volume [3]. Mandelbrot defined a time-scale, called trading time which - compared to clock time - flows slowly during some period and fast during others [9][10]. Let t be clock time and $T(t)$ a trading time, typically multifractal time. Fractional Brownian motions in multifractal time is represented as $X(t) = B_H(T(t))$ where B_H denotes fractal Brownian motions. Clark analyzed the price series as a probability process $X(T(t))$ where directing process $T(t)$ is another probability process and proposed to apply cumulated trading volume as the directing process [2]. In this article, another time scale which evolves the price fluctuation is defined and the relations of the three time scales: real time, tick time and newly defined C-fluctuation time, in foreign exchange markets are analyzed. Takayasu et. al. showed that the statistics of trade intervals in foreign exchange data is well characterized by a Poisson process with its mean given by a moving average of trade intervals for about 2 minutes [11]. In this study the intervals between extreme values determined by a resolution scale are investigated. The data set used is trading prices of foreign exchange rates; US dollar (USD)/Japanese yen (JPY), USD/Euro (EUR), and EUR/JPY. The accuracy of the data is one minute and data within a minute are recorded in order of transaction.

2. Scaling of fluctuations

Given a time series g(t), local maxima determined by a resolution scale C along price axis are defined as follows [6]. If there exist $d_1, d_2 > 0$ such that $g(s-d_1) < g(s)-C$, $g(s+d_2) < g(s)-C$, and g attains its maximum at s in $[s-d_1, s+d_2]$, then $g(t)$ has a C-local maximum at s. The scale C is the accuracy of measuring the extreme values. Making C larger filters out more data. The ordinary local maximum is the case $C=0$: 0-local maximum. In Fig. 1, $g(t)$ has a C-local maximum at s_1^C, and C'-local maxima at $s_1^{C'} = s_1^C$ and $s_3^{C'}$. Likewise, if there exist $d_1, d_2 > 0$ such that $g(s-d_1) > g(s) - C$, $g(s+d_2) > g(s) - C$, and $g(s) = \min_{s-d_1<t<s+d_2} g(t)$, then $g(t)$ has a C-local minimum at s. In Fig. 2, $g(t)$ has a C-local minimum at s_2^C, and C'-local minima at $s_4^{C'} = s_2^C$ and $s_2^{C'}$. These extreme values are related to the ex-post optimal trading under the transaction cost C and based on the same idea as a charting technique: Kagi chart [8].

Given a resolution C, the time series $g(t)$ attains its C-extreme values (C-local maximum or C-local minimum) at $s_1^C, s_2^C, s_3^C, \cdots$. Let R(C) denotes the sum of the absolute variation between neighboring C-extreme values:

$$R(C) = \sum |g(s_{i+1}^C) - g(s_i^C)|. \tag{1}$$

The fold dimension is defined as follows. If the relation:

$$R(C)/C \sim C^{-D_f} \tag{2}$$

holds, we say that the exponent D_f is the fold dimension of the time series. Since $R(C)$ is an invariant for time scale stretching, the fold dimension is independent of time scale. Let D_b denote the local box dimension of a time series, and H the Hurst exponent [5]. The relation: $D_b = 2 - H$ holds. The fold dimension is a kind of latent dimension equal to $1/H$ in the self-affine function generated by a special case of Iterated Puncttion System [1],[7]. To correspond the fractal dimension of time series, $2-1/D_f$ is used. Given resolution C, a time scale: C-fluctuation time can be defined as

$$\tau_C(t) = \sharp(\{s_i^C | s_i^C < t\}) \tag{3}$$

where $\sharp()$ denotes counting measure. To calculate the local box dimension of the time series by Higuchi method, the graph of the time series is covered with rectangles whose base are equal in length [4]. Here we cover the graph with rectangles equal in height. Applying C-fluctuation time as the time scale, these rectangles are also equal in length, and therefore the fold dimension is calculated by the procedure same as Higuchi method. The relation between C-fluctuation time and real (or tick) time The time intervals between these neighboring C-extreme values can be measured by real-time (clock time) and tick-time (the order of submission)

3. Results

The data used is foreign exchange rates tick data of US dollar (USD)/Japanese yen (JPY), USD/Euro (EUR), and EUR/JPY provided by CQG Inc. The record is given in the order of transaction. The transaction time is recorded with an accuracy of one minute. The data within a minute are recorded in order of transaction. There are four types of transaction; bid, ask, trade and settle, and the study presented here uses trading price.

Four periods in 2008 are analyzed in this article: July 14-18, August 18-22, September 15-19, and October 27-31. Each period is five business days of one week since Monday until Friday. Fig. 3 shows the distribution of the fluctuations: the frequency of price changes from the last trade of USD/EUR, USD/JPY and EUR/JPY in these four weeks. In the first two periods the market was calm, and in the latter two periods the market was volatile.

Fig. 6 shows double logarithmic plot of $R(C)/C$ with C of EUR/USD in the four periods. The fold dimension D_f can be calculated from the effective slope of this double logarithmic plot. The fold dimension D_f of each period is calculated with the scaling region of 5-20 pips. To compare the fractal dimension of time series, $2 - 1/D_f$ is shown in Fig. 7. The exchange rates are persistent in Sep. 15-19 and Oct. 27-31, especially in EUR/JPY.

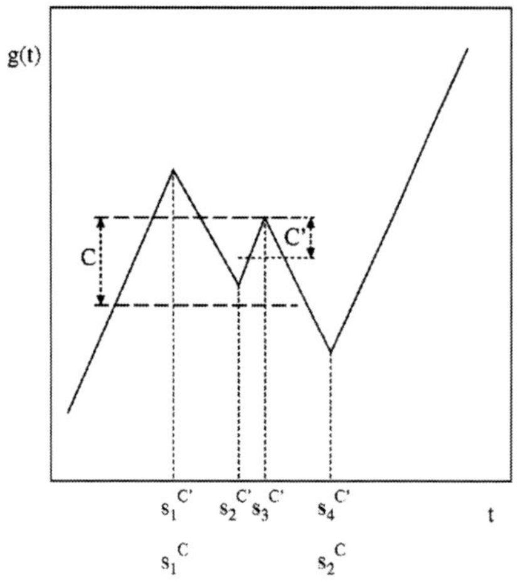

Figure 1. C-local maximum. $g(t)$ has a C-local maximum at s_1^C, and C'-local maxima at $s_1^{C'} = s_1^C$ and $s_3^{C'}$.

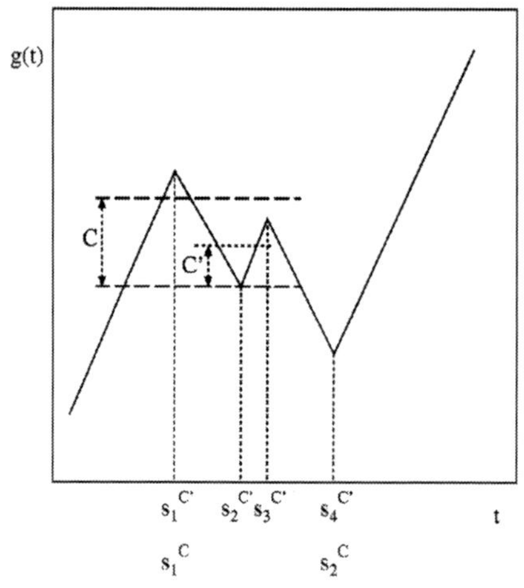

Figure 2. C-local minimum. $g(t)$ has a C-local minimum at s_2^C, and C'-local minima at $s_4^{C'} = s_2^C$ and $s_2^{C'}$.

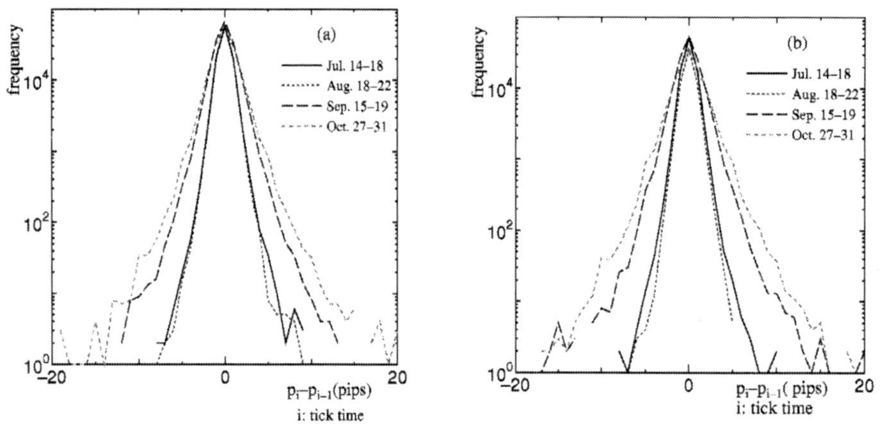

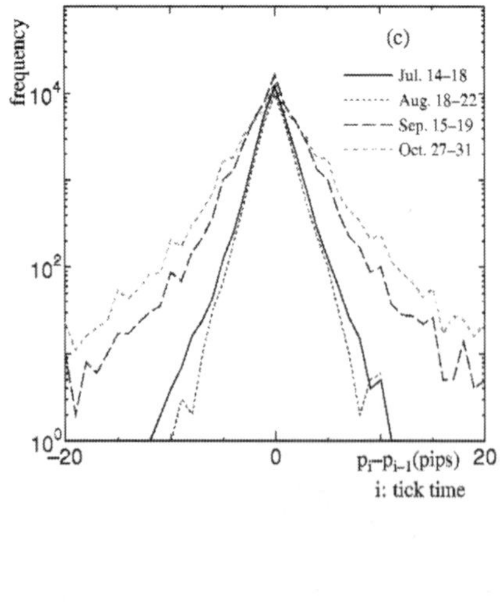

Figure 3. The distribution of the fluctuations: the frequency of price changes from the last trade; Jul. 14-18, Aug. 18-22, Sep. 15-19 and Oct. 27-31, 2008. (a)USD/EUR, (b)USD/JPY, (c)EUR/JPY

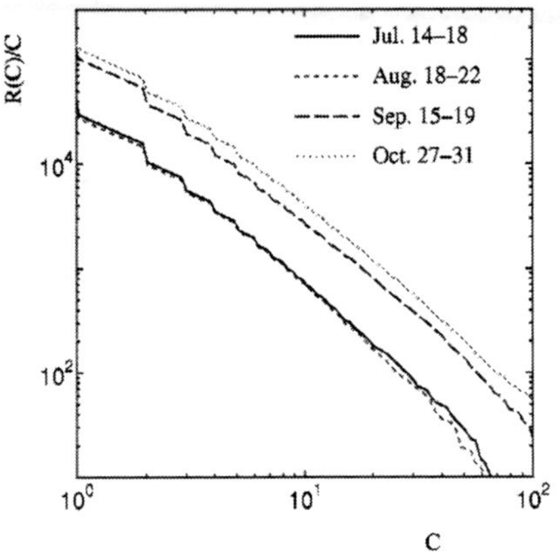

Figure 4. Double logarithmic plot of $R(C)/C$ with C (EUR/USD); Jul. 14-18, Aug. 18-22, Sep. 15-19 and Oct. 27-31, 2008.

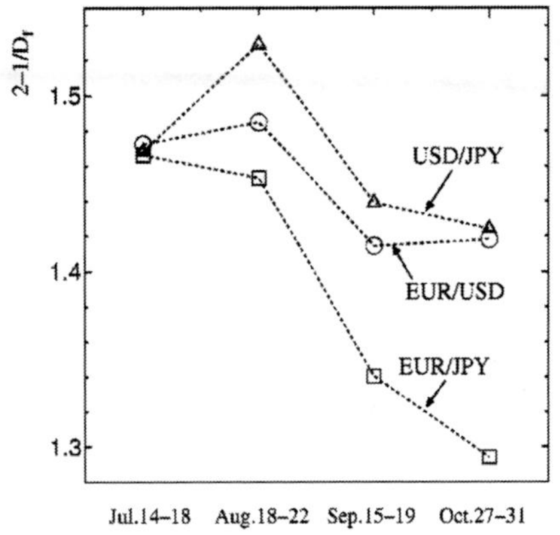

Figure 5. $(2 - 1/D_f)$ for EUR/USD (○), USD/JPY (△) and EUR/JPY (□) in Jul. 14-18, Aug. 18-22, Sep. 15-19 and Oct. 27-31, 2008. D_f is the fold dimension calculated with scaling region 5-20 pips.

Fig. 6 and 7 show examples of the relation of the three time scales: real time, tick time and C-fluctuation time (C=7pips) in the case of EUR/USD, Oct. 29, 2008. Fig. 6 shows the transformation of the time scale: C-fluctuation time into real time. In Fig. 7 a part of the transformation is from C-fluctuation time into tick time.

Fig. 8-13 show the distribution of the intervals of C-extreme values. In Fig. 8-10 the intervals between C-extreme values are measured by tick time. In Fig. 11-13 the intervals are measured by real time (minutes). Since time is recorded with an accuracy of one minute, the intervals are also measured by real time with an accuracy of one minute. The interval of 0 minute means that the interval is less than 60 seconds and the measuring error is greater than 0 second and less than 60 seconds. In case of the interval of other minutes, the measuring error is greater than -60 seconds and less than 60 seconds. Considering this measuring error, the bin width of 0 minute must be a half of these of other minutes. The figures (a) of Fig. 8-13 are the distributions in Aug. 18-22 and The figures (b) of Fig. 8-13 are the ones in Sep. 15-19. Fig. 8 and 11 show the distribution of EUR/USD, Fig. 9 and 12 show the ones of USD/JPY and Fig. 10 and 13 show the ones of EUR/JPY. The interval between C-extreme values is the increment of C-fluctuation time measured by another time scale such as real time. With small resolution C (1-2 pips), the intervals between C-extreme values are exponentially distributed in tick time. In EUR/JPY the intervals between C-extreme values are exponentially distributed with larger range. When the market is volatile, for larger resolution C (1-4 pips), the intervals between C-extreme values are exponentially distributed. When the market is not volatile, with the resolution C (3-5 pips), the intervals between C-extreme values are exponentially distributed in real time. When the market is volatile, for larger resolution C (5-15 pips), the intervals between C-extreme values are exponentially distributed in real time.

4. Summary

The series of instantaneous velocity of C-fluctuation time flowing, i.e. the intervals between C-extreme values, are exponentially distributed for small C when they are measured by real time and for smaller C (1-2 pips) when they are measured by tick time. When the market is volatile,

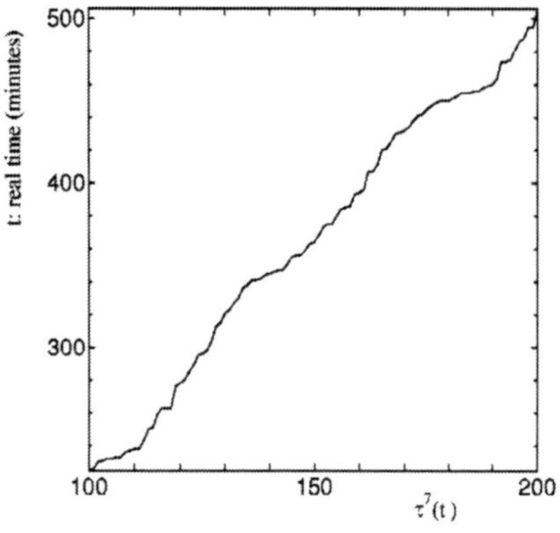

Figure 6. transformation of the time scale: C-fluctuation time (C=7pips) into real time (EUR/USD, Oct. 29, 2008).

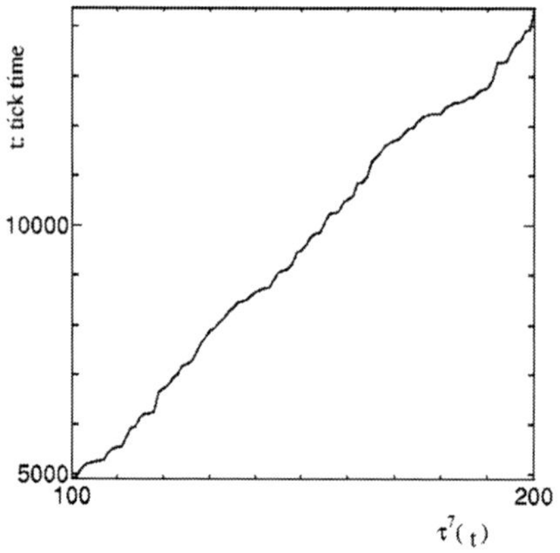

Figure 7. transformation of the time scale: C-fluctuation time (C=7pips) into tick time (EUR/USD, Oct. 29, 2008).

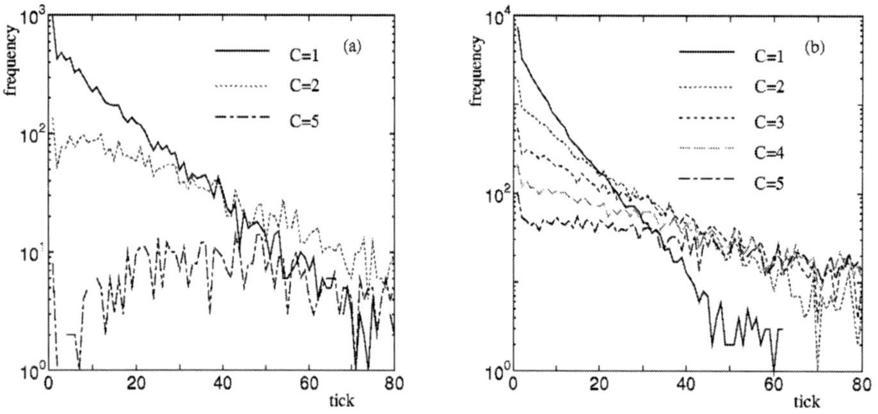

Figure 8. Interval of C-extreme values measured by tick time (EUR/USD, 2008) (a)Aug.18-22, (b) Sep.15-19

for larger C (1-4 pips), the intervals between C-extreme values are exponentially distributed. When the market is not volatile, with the resolution C (3-5 pips), the intervals between C-extreme values are exponentially distributed in real time. When the market is volatile, for larger resolution C (5-15 pips), the intervals between C-extreme values are exponentially distributed in real time.

Acknowledgments

The author acknowledges the continuing guidance and encouragement of Gyoichi Iwata, Toshio Inaba and Yukitami Tsuji.

References

[1] Barnsley M and Demko S 1985 *Proc. R. Soc. Lond.* A **399** 243-75

[2] Clark P K 1973 *Econometrica* **41** 135-55

[3] Ghysels E 2000 *J. Business and Economics Stat.* **18** 154-63

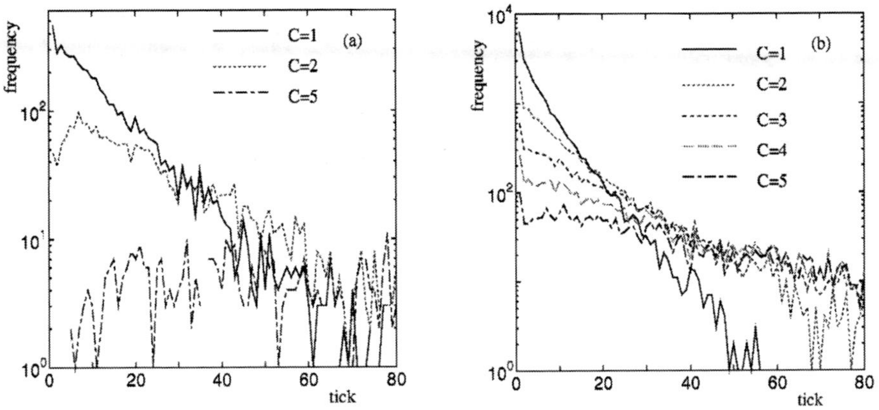

Figure 9. Interval of C-extreme values measured by tick time (USD/JPY, 2008) (a)Aug.18-22, (b) Sep.15-19

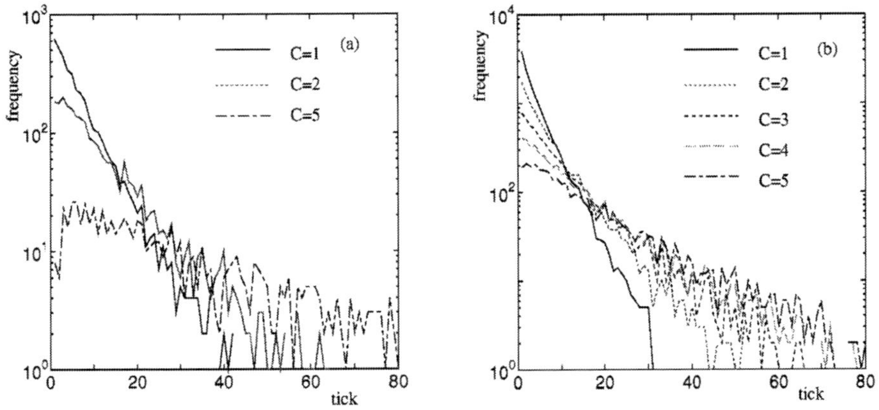

Figure 10. Interval of C-extreme values measured by tick time (EUR/JPY, 2008) (a)Aug.18-22, (b) Sep.15-19

[4] Higuchi T 1988 *Physica* D **31** 277
[5] Hurst H E 1950 *Trans. American Soc. Civil Engineers* **116** 770-808
[6] Kumagai Y 2002 Time-space scaling of financial time series *Empirical Science of financial fluctuations: The Advent of Econophysics* ed H Takayasu (Tokyo: Springer-Verlag) pp 250-259
[7] Kumagai Y 2002 *Fractals* **10** 13-8
[8] Kumagai Y 2003 Time evolution of fractal structure by price-axis scaling and foreign exchange intervention operations *The Application of Econophysics - Proc. of the 2nd Nikkei Econophysics Symp.* ed H Takayasu, (Tokyo: Springer-Verlag) pp 58-63
[9] Mandelbrot B B 1967 *J. Business* **40** 393-413
[10] Mandelbrot B B 1997 *Fractals and Scaling in Finance: Discontinuity, Concentration, Risk* (New York: Springer-Verlag)
[11] Takayasu M, Takayasu H and Ocazaki M P 2002 Transaction Interval Analysis of High Resolution Foreign Exchange Data *Empirical Science of financial fluctuations: The Advent of Econophysics* ed H Takayasu (Tokyo: Springer-Verlag) pp 18-25

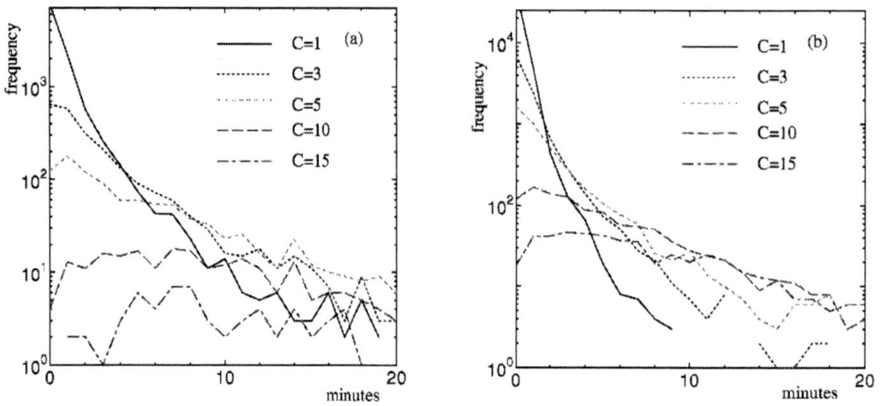

Figure 11. Interval of C-extreme values measured by real time (EUR/USD, 2008) (a)Aug.18-22, (b) Sep.15-19

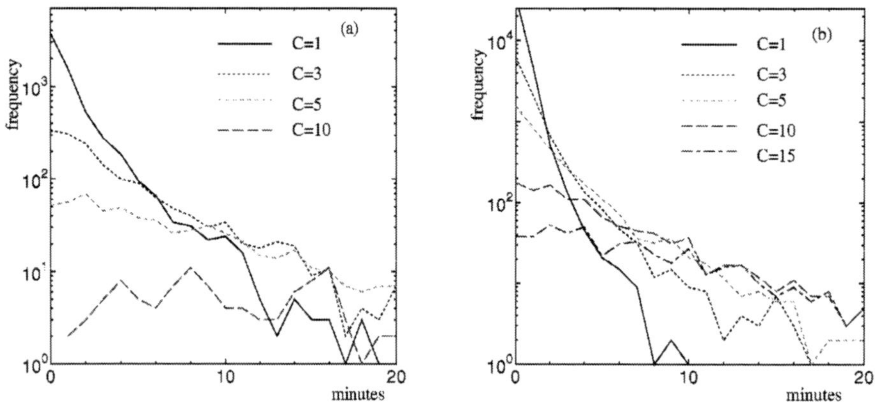

Figure 12. Interval of C-extreme values measured by real time (USD/JPY, 2008) (a)Aug.18-22, (b) Sep.15-19

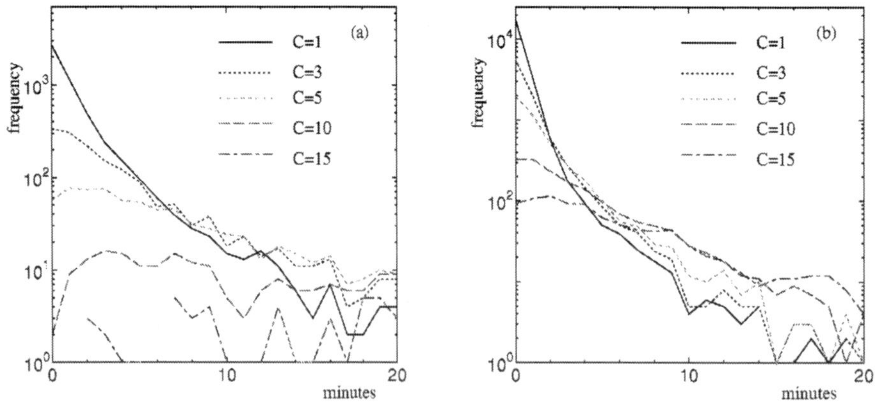

Figure 13. Interval of C-extreme values measured by real time (EUR/JPY, 2008) (a)Aug.18-22, (b) Sep.15-19

Forecasting the Portuguese Stock Market Time Series by Using Artificial Neural Networks

Monica Isfan[1,2], **Rui Menezes**[3] **and Diana A. Mendes**[3]

[1]Department of Statistical Metodology, INE, Avenida António José de Almeida, 1000-043 Lisbon, Portugal
[2]ISCTE-IUL, Avenida das Forças Armadas, 1649-026 Lisbon, Portugal
[3]Department of Quantitative Methods, ISCTE-IUL and UNIDE, Avenida das Forças Armadas, 1649-026 Lisbon, Portugal

E-mail: monica.isfan@ine.pt; rui.menezes@iscte.pt and diana.mendes@iscte.pt

Abstract. In this paper, we show that neural networks can be used to uncover the non-linearity that exists in the financial data. First, we follow a traditional approach by analysing the deterministic/stochastic characteristics of the Portuguese stock market data and some typical features are studied, like the Hurst exponents, among others. We also simulate a BDS test to investigate nonlinearities and the results are as expected: the financial time series do not exhibit linear dependence. Secondly, we trained four types of neural networks for the stock markets and used the models to make forecasts. The artificial neural networks were obtained using a three-layer feed-forward topology and the back-propagation learning algorithm. The quite large number of parameters that must be selected to develop a neural network forecasting model involves some trial and as a consequence the error is not small enough. In order to improve this we use a nonlinear optimization algorithm to minimize the error. Finally, the output of the 4 models is quite similar, leading to a qualitative forecast that we compare with the results of the application of *k*-nearest-neighbor for the same time series.

1. Introduction

Time series analyses and in particular forecasting of financial data have attracted some special attention in the last years. Unfortunately, the nonlinear nature and the complex behaviour of this type of data transform forecasting into a very hard task, and the classical statistical methods turn to be not adequate. A new generation of methodologies, including artificial neural networks (ANN) and genetic algorithms (GA) has been used for the analysis of trends and patterns, classification and forecasting. In particular, the use of computational intelligence based techniques for forecasting has proved to be extremely successful in recent times due to the ability of neural networks to approximate nonlinear functions [2, 8, 13, 14, 16, 20, 21].

Stock market prediction is very difficult. The efficient market hypothesis states that the current market price fully reflects all available information. This implies that past and current information is incorporated into stock price, thus price changes are merely due to new information and irrelevant to existing information. Since news occur randomly and cannot be forecasted, stock price should follow a random walk pattern. If this hypothesis is true, then any attempts to predict market will be futile. The random walk model has been tested extensively in various markets. The results are mixed and

sometimes contradictory. However, most of the recent studies [6, 7, 8, 10, 16, 17] on stock markets reject the random walk behavior of stock prices. Although numerous reports produce evidence that stock prices are not purely random, they all agree that the behavior of stock prices is approximately close to a random walk process.
Moreover, if the underlying process to a market time series is a fractional random walk, which can be deduced by estimating the Hurst exponent, then the only methods available for prediction are the neural algorithms [16, 18].

An artificial neural network is a mathematical model or a computational model based on biological neural networks or, in other words, is an emulation of a biological neural system. It consists of an interconnected group of artificial neurons and processes information using a connectionist approach to computation. The ANN is a powerful tool for nonlinear time series models and, in the last years, was used with success in solving nonlinear prediction and forecasting problems. In particular, ANNs are used to forecast financial markets, since they are able to learn nonlinear mappings between inputs and outputs, the systems' model is no needed (no priori assumption is needed) and can be applied to no-stationary data. Neural network forecasting models have been widely used in financial time series analysis during the last decade [2, 7, 13, 14, 15, 18, 20, 21].

In this paper, we show that neural networks can be used to uncover the non-linearity that exists in the financial field. First, we follow a traditional approach by analysing some of the deterministic/stochastic characteristics of the Portuguese stock exchange index and some typical features are done. We also produce a BDS (Brock, Dechert, and Scheinckman, 1987) test to investigate the nonlinearity and the result was as expected: the financial time series do not exhibit linear dependence. We analyze several periods with large and small Hurst exponents in order to obtain indications to make efficient the process of prediction time series with a multi-layer ANN with supervised learning. Secondly, we trained four types of neural networks for the stock markets and used the models to make a forecast.

ANN models were obtained using a three-layer feed-forward topology and the back-propagation learning optimization algorithm. The quite large number of parameters that must be selected to develop a neural network forecasting model involves some trial and as a consequence the error is not small enough. In order to improve this we use some nonlinear optimization numerical algorithm to minimize the error. Finally, the output of the 4 models was quite similar, leading to a qualitative forecast.

To compare the forecast errors for the above considered cases; we also forecast the data by using nearest neighborhood methods. By analyzing the response of various configurations to our data series, with specific characteristics; we conclude that the forecasting results when neural networks are applied are better than k-nearest-neighbor forecast results.

Nearest neighbor forecasting models are attractive with their simplicity and the ability to predict complex nonlinear behavior. They rely on the assumption that observations similar to the target one are also likely to have similar outcomes. A common practice in nearest neighbor model selection is to compute the globally optimal number of neighbors on a validation set, which is later applied for all incoming queries. For certain queries, however, this number may be suboptimal and forecasts that deviate a lot from the true realization could be produced.

2. Methodology

Formally, the problem of prediction of future values of some time series $y(t)$, $t = 1,2,\ldots$, can be formulated as finding a (nonlinear) function F so as to obtain an estimate $y'(t + D)$ of the vector y at time $t+D$ ($D = 1,2, \ldots$), given the values of y up to time t, plus a number of additional time independent variables (exogenous features) u_i:

$$y'(t+D) = F(y(t), \ldots y(t-dy), u(t), \ldots, u(t-du)),$$

where $u(t)$ and $y(t)$ represent the input and output of the model at time t, du and dy are the *lags* of the input and output of the system and F a nonlinear function. Typically, $D = 1$, means *one-step ahead*, but we can take any value larger than 1 and we say *multi-step ahead* prediction.
Viewed this way, prediction becomes a problem of (nonlinear) function approximation, where the purpose of the method is to approximate the continuous function F as closely as possible.

Usually, the evaluation of prediction performance is done by computing an error measure E over a number of time series elements, such as a validation or test set:

$$E = \sum (y'(t-k)\, y(t-k)) \quad .$$

E is a function measuring the error between the estimated (predicted) and actual sequence element. Typically, a distance measure (Euclidean or other) is used, but depending on the problem, any function can be used.

The adopted methodology in this work is based on three main steps: firstly, we established the architecture for the ANN; in particular a multilayer feed-forward artificial neural network with supervised training realized through the back-propagation algorithm. The error function was optimized by a numerical optimization technique based on Levenberg-Marquard algorithm. Secondly, in order to test the forecasting ability we due some basic descriptive statistics of the data and estimate the Hurst exponents. Finally, we forecast the future values for the Portuguese stock exchange (index) time series by using ANN and nearest neighbour with correlation and with absolute distance.

2.1. Feed-forward Artificial Neural Networks

Feed-forward neural networks [13], [15] are the simplest form of ANN. A feed-forward neural network (figure 1) is a biologically inspired classification algorithm. It consists of a (possibly large) number of simple neuron-like processing units, organized in layers. Every unit in a layer is connected with all the units in the previous layer. These connections are not all equal; each connection may have a different strength or weight. The weights on these connections encode the knowledge of a network and the transformation of the node information is realized by the activation function.

Data enters at the inputs and passes through the network, layer by layer, until it arrives at the outputs. During normal operation, that is when it acts as a classifier, there is no feedback between layers. This is why they are called feed-forward neural networks.

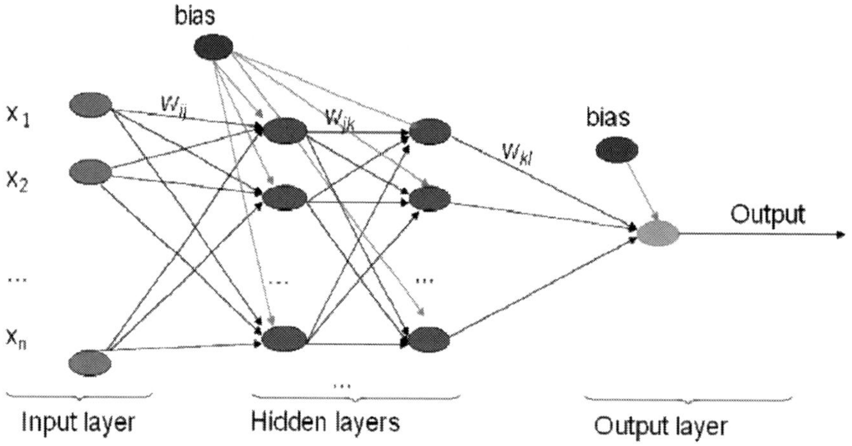

Figure 1. Feed-forward artificial neural network

Training an artificial neural network involves presenting input patterns in a way so that the system minimizes its errors and improves its performance. The training algorithm may vary depending on the network architecture, but the most common training algorithm used is back-propagation [6], [10]. This is the process of back-propagating errors through the system from the output layer towards the input layer during the training (figure 2). Back-propagation is necessary because hidden units have no training target value that can be used, so they must be trained based on errors from previous layers. The output layer is the only layer which has a target value which can be compared. As the errors are back-propagated through the nodes, the connection weights are changed. Training occurs until the errors in the weights are sufficiently small to be accepted.

Back-propagation algorithms are often too slow for practical problems, so we can use several high performance algorithms [9]. These faster algorithms fall into two main categories: heuristic techniques (as variables learning rate back-propagation, resilient back-propagation) and numerical optimization techniques (as conjugate gradient, Levenberg-Marquardt, etc.). In this paper, Levenberg-Marquardt algorithm, the sigmoid transfer function in the hidden layer and a linear transfer function in the output layer were suitable for our study.

The major problem in training an ANN is deciding when to stop training. Since the ability to generalize is fundamental for these networks to predict future values, overtraining is a serious problem. Overtraining occurs when the system memorize patterns and thus looses the ability to generalize. Overtraining can occur by having too many hidden nodes or training for too many time periods (epochs). However, overtraining can be prevented by performing test and train procedures or cross-validation. The test and train procedure involves training the network on most of the input data (around 90%) and testing on the remaining data. The network performance on the test set is a good indicator of its ability to generalize and handle data it has not been trained on. If the performance on the test is poor, the network configuration or learning parameters can be changed. The network is then retrained until its performance is satisfactory.

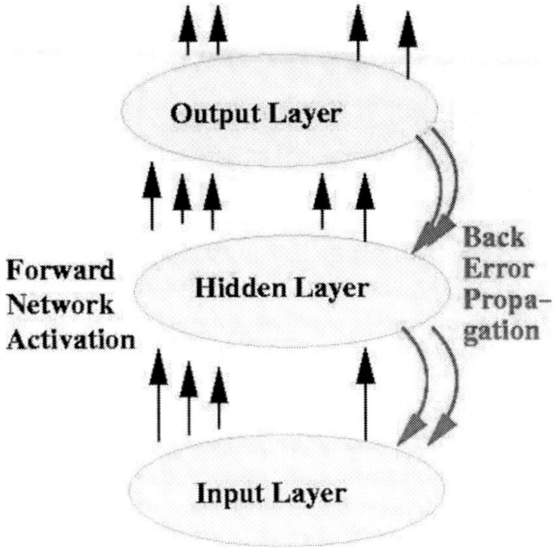

Figure 2. Backpropagation algorithm

2.2. Hurst exponent

In time series forecasting, the first question we want to know is whether the data under study is predictable. If the time series is random, then all methods are expected to fail. For this reason, we need some more details about the data, and we try to identify and study those time series having at least some degree of predictability.

The Hurst exponent, proposed by H. E. Hurst [10] for use in fractal analysis, has been applied to many research fields, and recently in finance domain. The Hurst exponent provides a measure for long-term memory and fractality of a time series and in consequence, can be used as a numerical estimate of the predictability of a time series. It is defined as the relative tendency of a time series to either regress to a longer term mean value or cluster in a direction. The reason the Hurst Exponent is an estimate and not a definitive measure is because the algorithm operates under the assumption that the time series is a pure fractal, which is not entirely true for most financial time series.

In conjunction with technical indicators or neural networks, Hurst Exponent estimation can help to determine which assets to forecast and which ones to ignore. This can be particularly useful in neural nets where models can focus more on time series with higher predictability. The date period is usually chosen arbitrary without any explanations. In this paper, we use the Hurst exponent to select period with great predictability to do further investigation.

The Hurst exponent [18] can be calculated by rescaled range analysis (R/S analysis). Hurst find that (R/S) scales by power-law as time increases, which indicates

$$(R/S)_t = ct^H \qquad (1)$$

where R is the range series, S is the standard deviation series, (R/S) is the rescaled range series, c is a constant and H is the Hurst exponent.

The values of the Hurst exponent range between 0 and 1. Based on the Hurst exponent value H, a time series can be classified into three categories:

- $H=0.5$ indicates a random series;
- $0<H<0.5$ indicates an anti-persistent series;

- $0.5<H<1$ indicates a persistent series.

An anti-persistent series has characteristics of "mean-reverting", which means an up value is more likely followed by a down value, and vice-versa. The strength of "mean-reverting" increases as H approaches 0.0. A persistent series is trend reinforcing, which means the direction of the next value is more likely the same as current value. The strength of trend increases as H approaches 1.0. Most economic and financial series are persistent with $H>0.5$. To estimate the Hurst exponent, we plot (R/S) versus t in log-log axes. The slope of the regression line approximates the Hurst exponent.

2.3. Nearest neighbour algorithm

The k-nearest neighbour (k-NN) algorithm is defined as a non-parametric regression. In order to obtain some forecast, we first have to identify the k most similar time series to a given query and then, by combining their historical continuations, evaluate the expected outcome for the query. For the k-NN the location of the observation in time is not important, because the objective of the algorithm is to locate similar pieces of information, independently of the respective time location. Thus the main idea of k-NN is to capture a non-linear dynamic of self similarity of the series.

The methods are linear with respect to the model parameters, and yet they turn out to be suitable for predicting highly nonlinear fluctuations too. This is due to the fact that the identified neighbors themselves could comprise complex nonlinear patterns.

The univariate nearest neighbour correlation method. The univariate case for k-NN algorithm [4, 5] works with the following steps:

- We define a starting training period and divide the period on different sectors or pieces y_t^m of size m, where $t=m,...,T$ and T is the number of observation on the training period. The term m is also defined as the embedding dimension of the time series. The last sector available before the observation to be forecast will be called y_T^m, and the other pieces will be called y_i^m.
- We select k pieces most similar to y_T^m. For the method of correlation, that means that we search for the k pieces with the highest value of $|q|$, which represents the absolute correlation between y_i^m and y_T^m.
- With k pieces on hand (each one with m observations), is necessary to understand in which way the k vectors can be used to construct the forecast on $(t+1)$. Several methods can be considered here, including the use of an average or of a tricube function.

The univariate nearest neighbour absolute distance method. The method of absolute distance [4, 5] is much simpler than the method of correlation, and the steps to follow are:

- We define a starting training period and divide the period on different sectors or pieces y_t^m of size m, where $t=m,...,T$ and T is the number of observation on the training period. The term m is also defined as the embedding dimension of the time series. The last sector available before the observation to be forecast will be called y_T^m, and the other pieces will be called y_i^m.
- We select k pieces most similar to y_T^m. For the method of absolute distance, that means that we search for the k pieces with the lowest sum of distances between the vectors y_i^m and y_T^m.
- With k pieces on hand (each one with m observations), is necessary to understand in which way the k vectors can be used to construct the forecast on $(t+1)$. The absolute distance method simply verifies the observations ahead of the k chosen neighbours and takes the average of them.

3. Data

The input variables are given by the Portuguese stock exchange market and consist of 4789 daily observations; 5 days per week (see figure 3). The data is non-stationary, with high skewness and kurtosis, and non-normal distribution, as can be observed from figure 4.

The BDS statistics clearly shows that there is no linear dependence in the data (see table 1). The BDS test tests the null hypothesis of independent and identical distribution (i.i.d.) in the data against an unspecified departure from i.i.d. A rejection of the i.i.d. null hypothesis in the BDS test is consistent with some type of dependence in the data, which could result from a linear stochastic system, a nonlinear stochastic system, or a nonlinear deterministic system. Under the null hypothesis, the BDS test statistic asymptotically converges to a standard normal distribution.

In table 1, $C(m,n)$ corresponds to the correlation integral, m is the embedding dimension and n the distance between the standard deviation of the data.

We applied the BDS test to our data for embedding dimensions of m = 2,3,4,5 and 6. We use the quantiles from the small sample simulations reported by Brock et al. [1] as approximations to the finite-sample critical values of our BDS statistics. The i.i.d. null hypothesis is rejected in all cases for yield changes.

The data were divided into two sets, the training set and the test set (3831 observation were used for the training of the network while the rest of 958 data was used for testing). The Hurst exponent, ANN models and k-NN prediction were analyzed and estimate by using MATLAB packages and programs.

The out-of-sample forecasting horizon is one-step ahead, with the competing forecasting models estimated over the training set and applied over the test set to generate genuine one-step-ahead out-of-sample forecasts. The criterion for forecasting performance is absolute mean.

The analysis was initially done for the original non-stationary data and after extended to the log difference data. We stress out that all the initial parameters and the algorithms implemented are the same for the original data and log difference data.

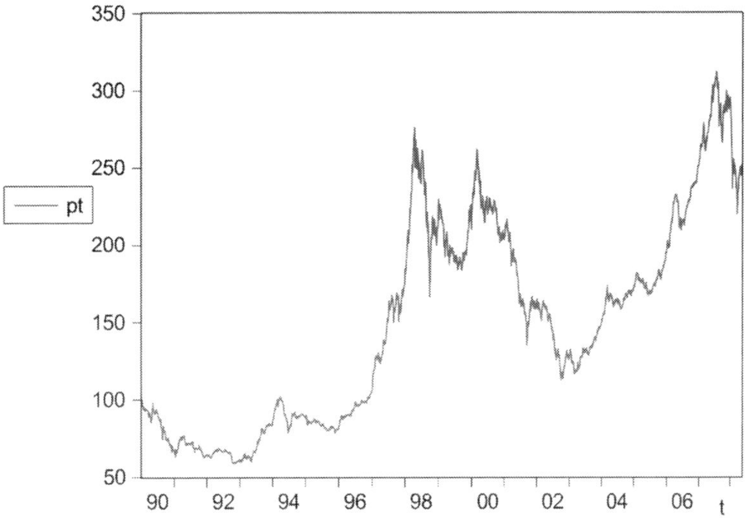

Figure 3. DataStream: PORTUGAL - DS Market

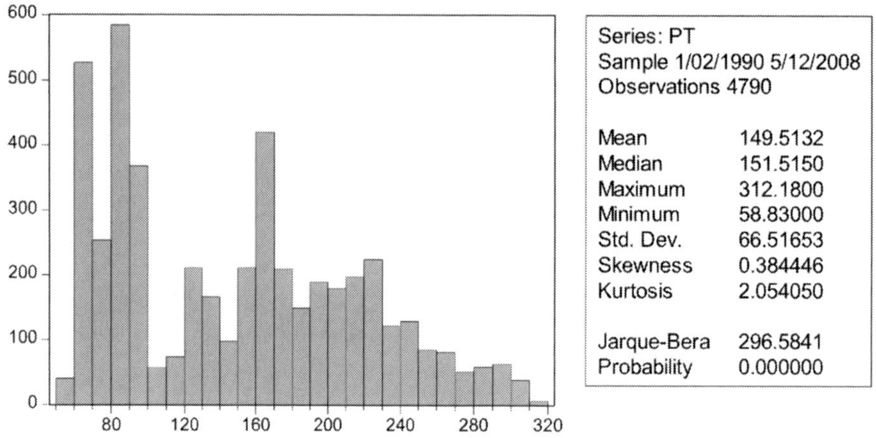

Figure 4. DataStream: PORTUGAL - DS Market

Table 1. BDS statistics for Portuguese Stock Market

BDS Test for PT
Date: 02/24/09 Time: 15:12
Sample: 1/02/1990 5/12/2008
Included observations: 4790

Dimension	BDS Statistic	Std. Error	z-Statistic	Prob.
2	0.204947	0.000698	293.7345	0.0000
3	0.348908	0.001103	316.2462	0.0000
4	0.449787	0.001306	344.2839	0.0000
5	0.520359	0.001354	384.3905	0.0000
6	0.569610	0.001298	438.9545	0.0000

Raw epsilon	103.0320		
Pairs within epsilon	16144790	V-Statistic	0.703658
Triples within epsilon	5.71E+10	V-Statistic	0.519276

Dimension	C(m,n)	c(m,n)	C(1,n-(m-1))	c(1,n-(m-1))	c(1,n-(m-1))^k
2	8026628.	0.700107	8067551.	0.703676	0.495160
3	7992952.	0.697461	8065114.	0.703757	0.348553
4	7963700.	0.695198	8062685.	0.703839	0.245411
5	7937382.	0.693191	8060263.	0.703922	0.172831
6	7913042.	0.691354	8057816.	0.704003	0.121743

4. Results

We do several tests in order to estimate the Hurst exponent, for the considered data. As we can observe from figure 5 and table 2 the average Hurst exponent we found is approximately equal to 0.65, meaning that the time series is trend reinforcing. This indicates that we are in the presence of a long memory dependence process with low fractal dimensions and in consequence the date is quite predictable.

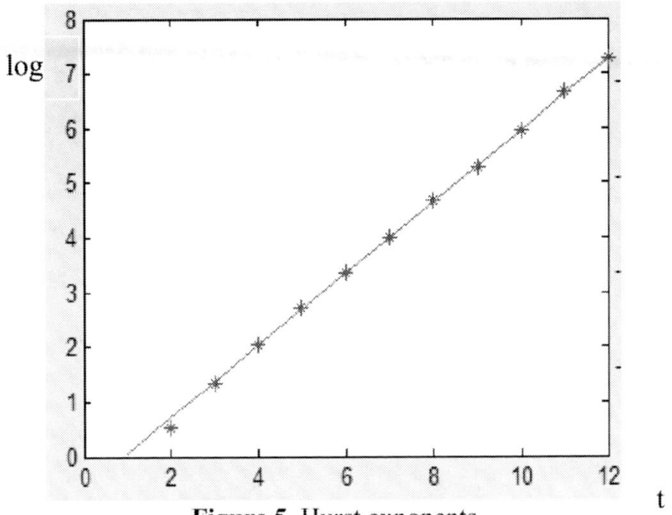

Figure 5. Hurst exponents

Table 2. Hurst exponents

Hurst exponents	
0.65692	(3:12)
0.65232	(4:12)
0.65153	(5:12)
0.65692	(3:12)
0.65765	(3:11)
0.6559	(3:10)
Hurst exponent - average	
0.65521	

In all the performed simulations and estimations, a one-step-ahead prediction is considered; that is, the actual observed values of all lagged samples are used as inputs.

In figure 6 and figure 7, respectively we have represented the real no-stationary data versus the forecasted data by using nearest neighbour with correlation and with absolute distance methods and the log difference data versus the forecasted data using the same methods as above. For the two forecast algorithms (nearest neighbour with correlation method and absolute distance method) and for both original data and log difference data, we considered *4700* observations on the training period, *k= 30* most similar pieces and *m=4* the embedding dimension. The results for the absolute mean error and correlation mean error are showed in table 3.

Table 3. Absolute mean error and correlation mean error

	Mean absolute error	**Correlation mean error**
Real data	1.0245	0.7819
Log difference data	1.8993e-004	8.1925e-004

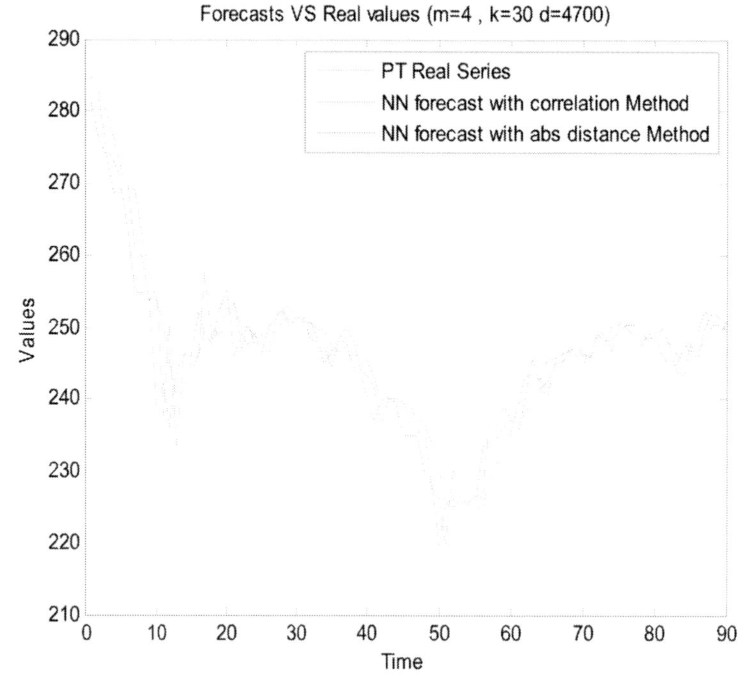

Figure 6. Forecast versus real data

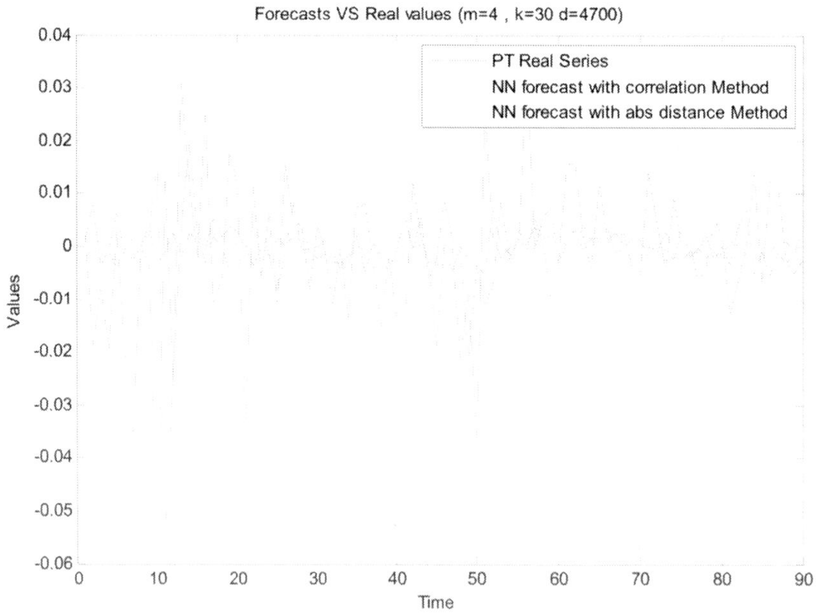

Figure 7. Forecast versus log difference data

We can observe that, for this forecasting method, we obtain better results if we apply the methods to the stationary log difference data, than for the non-stationary data.

In figure 8 and 9 we illustrate the forecast results for the original and respectively, the log difference data and the forecast of the error for the test set.

Various ANN models with different numbers of lagged input steps (or time windows), and different number of neurons in hidden layers have been compared. All the models had layer 1 (input) with N neurons and a single neuron in layer 2 (output layer). The main characteristics for our optimal ANN are registered in Table 4, namely the best performance was obtained for a 3-layer feed-forward neural network with 12 hidden units, low learning rate (0.1) and trained for maximum 2000 epochs. In order to obtain fast convergence we apply the Levenbeg-Marquardt algorithm.

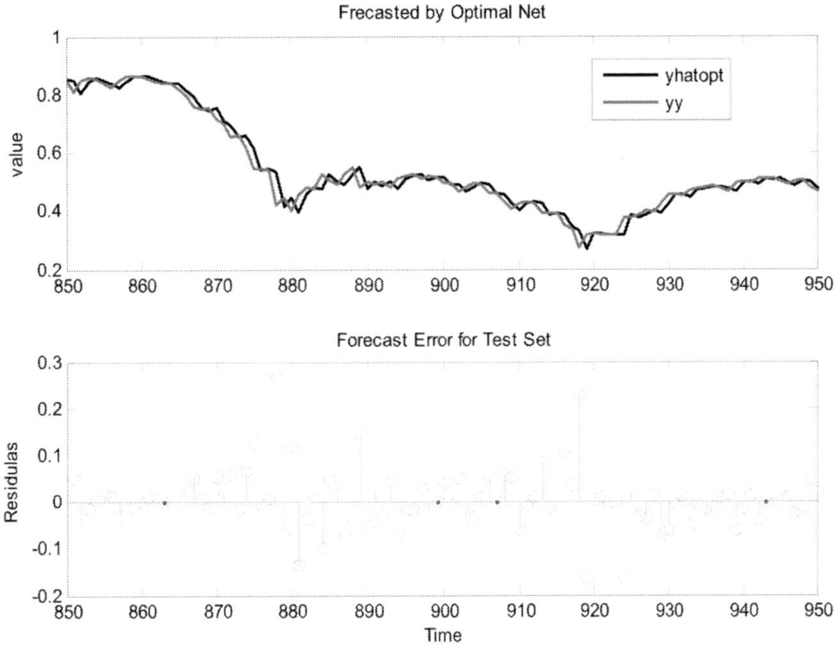

Figure 8. Forecast of original data by the optimal ANN

In figure 8 yy is the original data and yhatopt represents the forecast results for original data.
In figure 9, yy stands for the log difference data and yhatopt represents the forecast results for log difference data.

We can observe in this case that ANN methods perform a better prediction for the original no-stationary data than for the log difference data. We also highlight, that we obtain a lower forecast error for stationary log difference data by using k-nearest neighbour.

Consequently, the use of ANN provides fast convergence, high precision and strong forecasting ability of real data. Even if ANN's are not perfect in their prediction, they outperform all other methods and provide hope that in the future we can better understand the underlying dynamic of chaotic and/or stochastic systems such as the stock markets, between others.

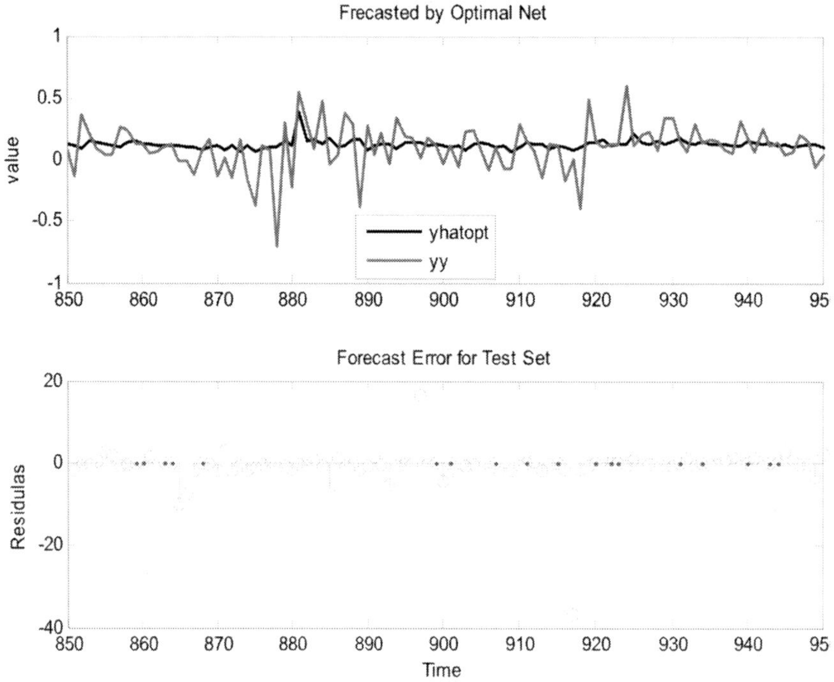

Figure 9. Forecast of log difference data by the optimal ANN

Table 4. Parameters of the optimal ANN for forecast

	No. of layers	**Hidden units**	**Training set**	**Output units**	**Learning rate**	**Training epochs**	**Min RMSE**
Original data	3	12	80%	5	0.1	2000	0.0066
Log differenc e data	5	10	80%	5	0.5	2000	0.0454

Acknowledgment: Financial support from the Fundação Ciência e Tecnologia, Lisbon, is grateful acknowledged by the authors, under the contracts No PTDC/GES/73418/2006 and No PTDC/GES/70529/2006.

References

[1] Brock, W., Dechert, W. D., and J. Scheinkman, 1987, A Test for Independence Based on the Correlation Dimension, *Economics Working Paper* SSRI-8702, University of Wisconsin.

[2] Enke, D. & Thawornwong, S., 2005, The use of data mining and neural networks for forecasting stock market returns, *Expert Systems with Applications*, 29 (4) 927-940.

[3] Farmer, D. and Sidorowich J, 1987, Predicting chaotic time series, *Physical review letters*, vol 59, p 845-848.

[4] Fernández-Rodríguez, F., Sosvilla-Riveiro, S. and Felix, J, A,, 2002, Nearest *neighbor predictions in foreign exchange markets* (Working paper), no. 05, FEDEA.

[5] Fernández-Rodríguez, F., Sosvilla-Riveiro, S. and García-Artile, M., 2001, *An empirical evaluation of non-linear trading rules* (Working paper), no.16, FEDEA.

[6] Gallagher, L. and M. Taylor, 2002, Permanent and Temporary Components of Stock Prices:

Evidence from Assessing Macroeconomic Stocks, *Southern Economic Journal*, 69, pp.245-262.

[7] Gately, E., 1996, *Neural networks for financial forecasting*, Wiley: New York, 1996.

[8] Gooijer, J.G. de & Hyndman, R.J., 2006, 25 years of time series forecasting. *International Journal of Forecasting*, 22, 443-473.

[9] Hayati, M. and Shirvany, Y., 2007, Artificial neural network approach forshort term load forecasting for Illam Region, *Proc. of World Academy of Science, Engineering ang Technology* (Volume 22, ISSN 1307-6884).

[10] Hurst, E. H ., 1951, Long-term storage of reservoires: an experimental study, *Transaction of the American society of civil engineering* (116, 770-799).

[11] Kavussanos, M.G. and E. Dockery, 2001, A multivariate test for stock market efficiency: the case of ASE, *Applied Financial Economics*, vol. 11, no. 5, pp. 573-579(7), 1 October 2001.

[12] Lawrence, R., 1997, Using neural networks to forecast stock market prices (Course project, University of Maritoba, Canada).

[13] McNelis, D. P., 2005, *Neural Networks in finance: gaining predictive edge in the market* (USA: Elsevier Academic Press).

[14] Moreno, D. & Olmeda, I. , 2007, The use of data mining and neural networks for forecasting stock market returns. *European Journal of Operational Research*.182 (1) 436-457.

[15] Priddy, L. K. and Keller, E. P., 2005, *Artificial Neural Network: an introduction* (Washington: Bellington) SPIE Press.

[16] Rotundo, G., Torozzi, B., and Valente, M., 1999, *Neural Networks for Financial Forecast*, private communication.

[17] Refenes, A., 1995, *Neural networks in the capital markets*, Wiley: New York.

[18] Qian, B. and Rasheed, K., 2004, Hurst exponent and financial market predictability, *Proc. of Financial Applications and Engineering*, 437.

[19] Yildiz, B., Yalama, A. and Coskum, M., 2008, Forecasting the Istambul stock exchange national 100 index using an artificial neural network, *Proc. of World Academy of Science, Engineering ang Technology* (Volume 36, ISSN 2070-3740).

[20] S. Walczak, S., 2001, An empirical analysis of data requirements for financial forecasting with neural networks, *Journal of Management Information Systems*, 17(4), pp.203-222.

[21] Zirilli, J.S., 1997, *Financial Prediction using Neural Networks*, International Thomson Computer Press: UK.

Life-span distributions of supermarket products

Hiromichi UENO, Tsutomu WATANABE and Misako TAKAYASU

Department of Computational Intelligence and Systems Science, Interdisciplinary Graduate School of Science and Engineering, Tokyo Institute of Technology, 4259-G3-52 Nagatsuta-cho, Midori-ku, Yokohama 226-8502, Japan and Institute of Economic Research, Hitotsubashi University 2-1 Naka, Kunitachi, Tokyo 186-8603, Japan

E-mail: ueno@smp.dis.titech.ac.jp

Abstract. We have analyzed the lifetime distributions of more than 0.7 million products sold across approximately 400 Japanese supermarkets. The distributions are well approximated by an exponential function for products with lifetimes longer than 1000 days, implying that the manufacturers' decisions about whether to continue production are purely random. However, the distributions tend to deviate from an exponential distribution for products with lifetimes shorter than 1000 days. Specifically, the distributions for food products exhibit a quicker decay in a short time scale, suggesting the existence of competing products during the initial stages of the product lifecycle. On the other hand, the distributions for toiletry products exhibit a slower decay in a short time scale.

1. Introduction

Developments in information technology have enabled the storage of large volumes of real-time economic data on computer systems. In the field of econophysics, such data are analyzed using statistical physics methods, and important results have already been established for financial markets and company's financial data[1][2]. Recently, data related to various types of human activities have been stored, and these have become the latest focus of attention in the field of econophysics. For example, Sornette et al. conducted an extensive study of the time series of books sales by using a unique database obtained from Amazon.com[3][4].

Recently, point of sale (POS) data has also been analyzed in this field. Most retailers employ a POS system for the purpose of efficient inventory management, accounting, and marketing. Such systems maintain a real-time record of the number of sales of products and the price of each. By using this system, retailers can monitor all purchasing activities in a retail store. Since POS data contains detailed information, it can be used to analyze economic activities from various points of view. Mizuno et al. studied the repeat purchase behavior of consumers by using POS data collected from Japanese convenience stores[5][6]. Groot analyzed the fluctuations in product sales using sales data collected from Dutch supermarkets and observed that these fluctuations exhibit properties similar to those of a stock market[7]. Around 30 years ago, Montroll analyzed the product pricing in the catalogue of Sears, Roebuck and Company and observed that the price distribution follows a log-normal distribution[8].

In this study, we analyze the product lifetime distributions using POS data collected from Japanese supermarkets. It is thought that life span characterizes the object's property. The lifetime distributions of various phenomena are typically represented by an exponential distribution or a Weibull distribution. For example, radioactive decay follows an exponential distribution; here, an element independently decays at a constant probability. On the other hand, the probability of fault occurrence in a machine, say an electron tube, follows the Weibull distribution[9].

We define the lifetime of a retail product as the period between the day when it first becomes available at a store (entry date) and the final day when it is sold (exit date). Manufacturers determine the entry and exit dates for a product according to their merchandising strategy. Therefore, the product lifetime influences the merchandising strategy and retail markets, and therefore, it is useful to analyze the product lifetime distribution. In the field of economics, the turnover of a product has been considered to be related to firms' technological innovation, in particular product innovation, which is a key element of firm and industry dynamics. The results of this paper will be useful in the given field[10][11].

2. Data

In this study, we analyze the POS data (provided by Nikkei Digital Media Inc.) collected from Japanese supermarkets. This data comprises many records, each of which consist of 6 numerical codes, namely, the JAN (Japanese Article Number) code (a unique product identification code), gencode (a supplementary code of JAN code), store code, date of a sale, gross sales price of the day, and number of sales in a day.

This set of codes is unique to each product. The POS data analyzed by us contains information about approximately 1.4 million types of products available across approximately 400 Japanese supermarkets between March 1988 and April 2008. We analyzed approximately 3.7 billion records.

This POS data includes information about all commercial products with a JAN+gencode that are sold at supermarkets; this enables a comprehensive analysis of product lifetimes. It should be noted that this data does not contain information abut products without JAN codes, including fresh foods such as vegetables, meat, and fish.

3. Definition of product lifetime

We define the product lifetime as follows. First, we define the entry date as the day when a product first becomes available in a store, and the exit date as the day when the product is sold. The sales lifetime is then defined as the number of days between the entry and exit dates. Fig.1 shows a schematic diagram that explains this definition. Here, the entry date is the date when the product first becomes available at store 2 (circle), and the exit date is the date when the product is sold in store 5 (square).

To define the product lifetime more explicitly, we excluded some products from the data analysis; these either had entry dates earlier than the beginning of our observation period or had exit dates later than the end of our observation period. Fig.2 shows a schematic diagram of the case in which the entry date of a product was earlier than the beginning of our observation period. The entry date of product B is represented by the circle; however, this date is the first day of our observation period, and the real entry date is earlier than this date. To avoid such estimation errors for both the entry and the exit dates, we apply the following rules. Products whose entry and exit dates were within 364 days of our first and last observation dates, respectively, were excluded from the analysis.

The threshold period was selected as 364 days because in some cases, a store or a manufacturer stopped selling a product for a certain period before resuming its sale. We observed the distribution of such intermissions and found that the probability of them lasting longer than a year was practically zero.

After filtering the products according to the above rules, approximately 0.7 million products remained, and we analyzed their lifetimes.

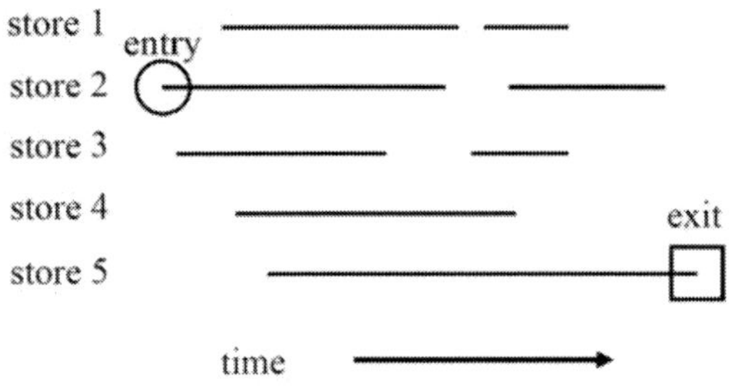

Figure 1. Schematic diagram explaining the definition of product lifetime on the basis of our sales data. The entry and exit dates are denoted by a circle and a square, respectively.

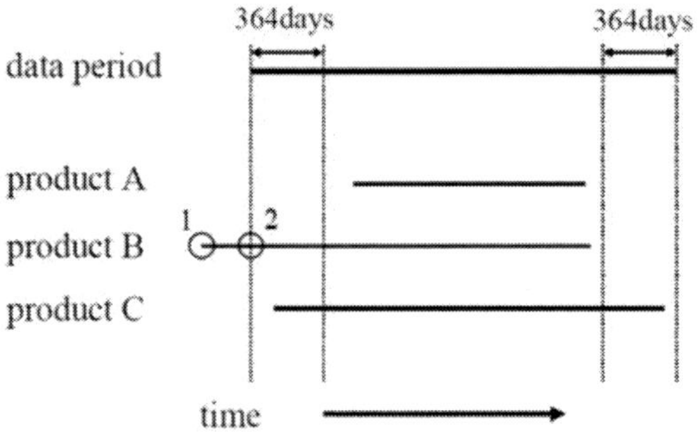

Figure 2. Schematic diagram explaining exclusion of a product having an entry date earlier than our observation period. The real entry date of product B (circle 1) does not coincide with the first day of the observation period.

4. Distribution of sales lifetime

We observe the cumulative distribution of the sales lifetime of all products that were analyzed, P(>=t), that is, the probability of finding a product having a lifetime greater than t days. Fig.3 shows a semilog plot of the probability distribution of products having P(>=t). It is evident that the curve can be approximated by a straight line except for the very short days and very long days. The dotted line represents the best-fit exponential function with an exponent of 0.001206, implying that the characteristic life span is 829 days (2.3 years).

The sharp decay in the distribution curve after around 6500 days (Fig.3) is caused by the limitation of the observation period to around 6500 days. If no such limitation is enforced, the exponential distribution is expected to extend longer.

To investigate the deviation from the exponential distribution in very short time scales, we plot the probability density function, that is, the derivative of -P(>=t), for a short time scale, as shown in Fig.4. We can confirm the presence of several peaks around the 31-, 62-, 92-, 123-, 154-, 183-, and 369-day periods. This implies that there is a tendency of making decision on determining the period of sales in the unit of month or year.

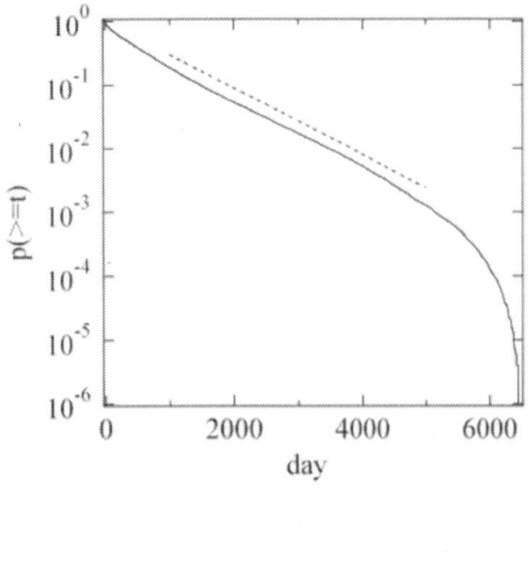

Figure 3. Semilog plot of cumulative distribution of sales lifetimes of all products in Japanese supermarkets. The dotted line represents an exponential function with a characteristic time of 829 days.

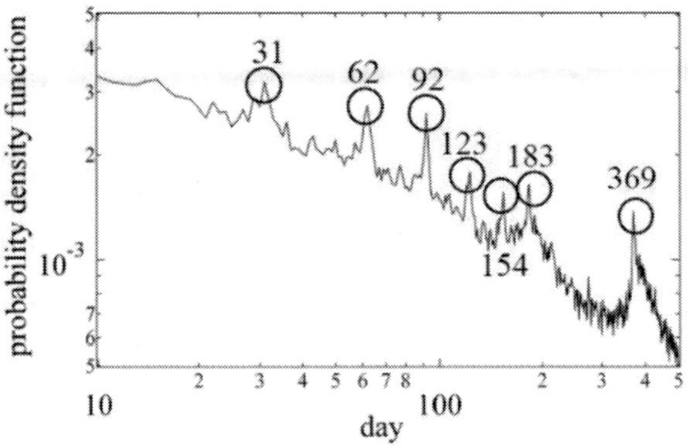

Figure 4. Probability density function of sales lifetimes of all products categories plotted on a log-log scale.

5. Dependence of distribution on differenct product categories

Products are classified into two broad categories, i.e., food products and toiletries, and these are further divided into 36 subcategories such as beverages, alcoholic drinks, detergents, etc. To investigate the heterogeneity of lifetimes in each category, we analyze the lifetime distributions of each category. Fig.5 shows the cumulative lifetime distribution for detergents (dotted line) and bread (solid line) on a semilog scale. The former roughly follows an exponential function over the entire time scale until the observation threshold, while the latter does not follow an exponential function over a time scale shorter than approximately 1000 days.

A short time scale is one that has probability values `P(>=t) > 0.5`, while a long time scale is one that has probability values in the range `0.1>=P(>=t)>0.01`. Fig.6 shows a graph of the combination of characteristic lifetimes for short and long time scales for all product categories. The horizontal (T1) and vertical (T2) axes represent the characteristic lifetimes for the short and long time scales, respectively. The plots on the guide line represent the case in which the two exponents coincide, i.e., when the distribution is approximated by one exponential function for the entire time scale.

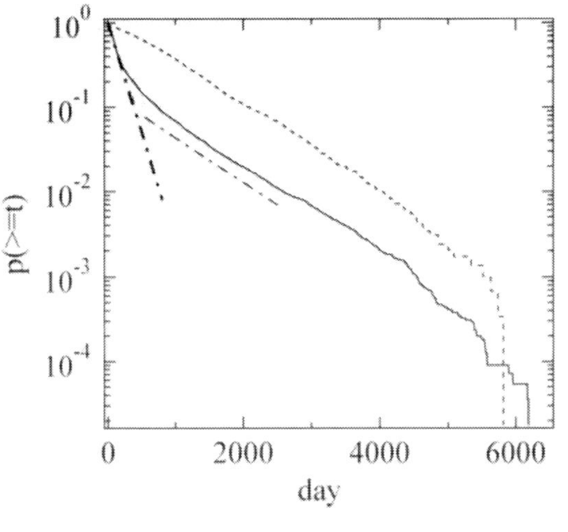

Figure 5. Distributions of sales lifetimes for the bread (solid line) and detergent (dotted line). The dashed line and bold dashed line represent the approximation by exponential functions.

The values for the categories of toiletries and food products are plotted on the lower-right and upper-left sides of the graph, respectively (Fig.6). The distributions are observed to be different for each category.

The characteristic lifetime of an exponential function depends on the product category. In

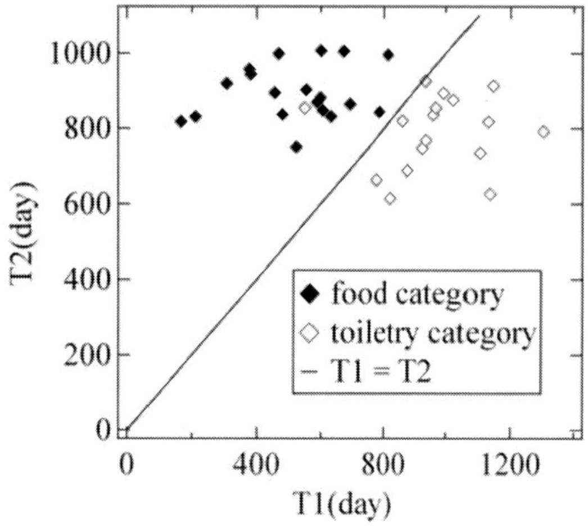

Figure 6. Combination of values of characteristic lifetimes for short and long time scales. The dotted line represents the case in which these values coincide. Products are divided into two broad categories, i.e., food products and toiletries.

the short time scale (T1), the average lifetimes lie in the range of 200-1300 days, as shown in Fig.6. For food products, the lifetimes are shorter than 800 days, while for toiletries, they are comparatively longer. In the long time scale, the characteristic life spans are generally around 800 days (2.2 years).

Fig.7 shows the lifetime distributions for 4 food products, namely, bread, chilled desert, terrine, and sweets. The distribution for each product is well approximated by the dotted straight line in the range of 2000-5000 days. In this time scale, the lifetime distributions for all food products roughly follow a universal exponential function. However, in the short time scale, the distribution exhibits a rapid decay.

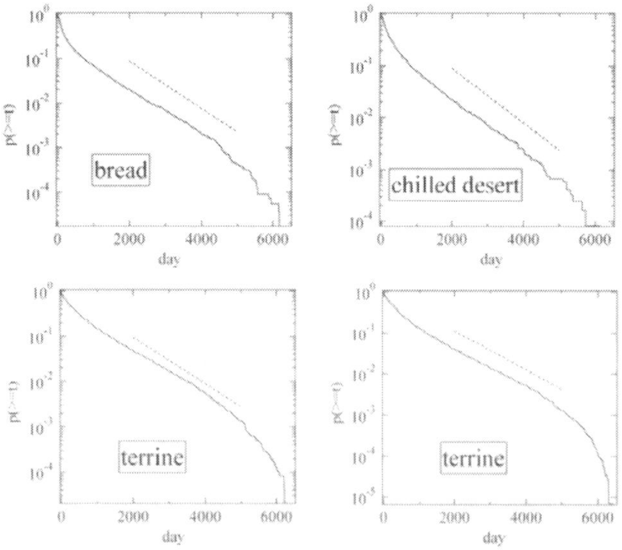

Figure 7. Cumulative lifetime distributions for 4 food products, namely, bread, chilled desert, terrine, and sweets.

Fig.8 shows the lifetime distributions for 4 toiletry products, namely, bath and body-care goods, sanitary goods, food and sanitary products for pets, and detergents. The distribution for each item is well approximated by the dotted line in the range of 2000-5000 days. However, in the short time scale, the distribution exhibits a slower decay.

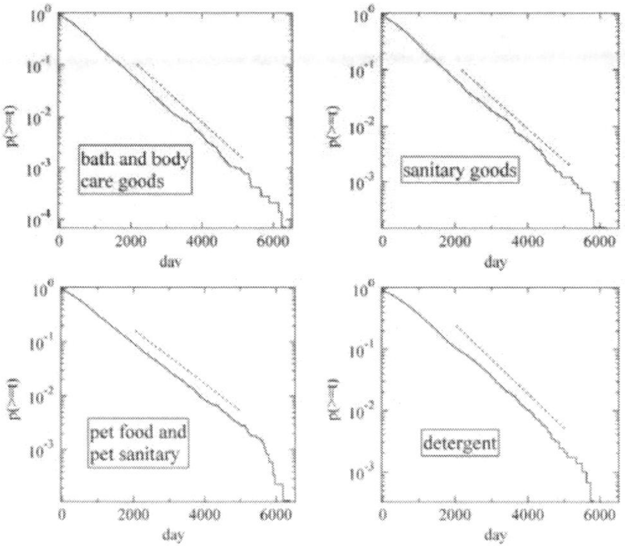

Figure 8. Cumulative lifetime distributions for 4 toiletry products, namely, bath and body-care goods, sanitary goods, food and sanitary products for pets, and detergents.

6. Summary

In this study, we observed that the probability density function of the sales lifetimes of supermarket products roughly follows an exponential function accompanied by several peaks around the unit of months or year. The distributions follow an exponential function rather than a Weibull distribution, implying that the manufacturers' decisions about whether to continue to produce or not can be well approximated by a random process such as a Poisson process; however, there is a tendency of making decision on determining the period of sales in the unit of month or year.

In a long time scale, the lifetime distributions of all product categories are well approximated by a common exponential function. This implies that in a large time scale, a product of each category exits nearly randomly at a constant rate.

However, in a short time scale, the distribution sales lifetimes is found to be dependent on the product category. In this time scale, the characteristic lifetimes of food products are generally short. This is because food products tend to exit at a rapid rate in a short time scale, possibly due to the existence of competing products. On the other hand, the characteristic lifetimes of toiletry products are generally long. This is because toiletry products tend to exit at a slow rate in a short time scale. The analysis of POS data revealed that in a short time scale the lifetime distributions for different product categories are different. It is believed that manufacturers implement different merchandising strategies for different product categories.

The POS data contains information about the number of sales of products and the price of each. By using a POS dataset, we intend to analyze the relation between a manufacturer's merchandising strategy and fluctuations in sales.

Acknowledgement

The authors appreciate the corporation of "Nikkei Digital Media Inc." for providing the POS data. This work is partly supported by the Research Fellowships of the Japan Society for the Promotion of Science.

References

[1] R.N.Mantegna and H.E.Stanly 1995 *Nature* **376** 46-49.
[2] M.H.R Stanly 1996 *Nature* **379** 804-806.

[3] D.Sornette, F.Deschatres, T.Gilbert, and Y.Ageon 2004 *PRL* **93** 228701.
[4] F.Deschatres and D.Sornette 2005 *PRE* **72** 016112.
[5] Takayuki Mizuno, Masahiro Toriyama, Takao Terano and Misako Takayasu 2008 *Physica A* **387** 3931-3935.
[6] Takayuki MIZUNO and Misako TAKAYASU 2009 *Progress of Theoretical Physics Supplement* **179** 71-79.
[7] Robert D. Groot 2005 *Physica A* **353** 501-514.
[8] ELLIOT W.MONTROLL 1981 *Proc. Natl. Acad. Sci. USA* **78-12** 7839-7843.
[9] John H. K. Kao 1959 *Technometrics* **1-4** 389-407.
[10] Christian Broda and David E. Weinstein 2007 *NBER Working paper* **13041**.
[11] Bils Mark 2004 *NBER Working paper* **10606**.

Simulating the Microstructure of Financial Markets

Tobias Preis[1234]

[1] Center for Polymer Studies, Department of Physics,
590 Commonwealth Avenue, Boston, Massachusetts 02215, USA
[2] Institute of Physics, Johannes Gutenberg University Mainz,
Staudinger Weg 7, 55128 Mainz, Germany
[3]Artemis Capital Asset Management GmbH,
Gartenstr. 14, 65558 Holzheim, Germany

E-mail: mail@tobiaspreis.de

Abstract. In the beginning of exchange based trading, floor trading was the most widespread form of trading. In the course of the introduction and the progress in information technology, trading processes were adapted to the computational infrastructure at the international financial markets and electronic exchanges were created. These fully electronic exchanges are the starting point for recent agent-based models in econophysics, in which the explicit structure of electronic order books is integrated. The electronic order book structure builds the underlying framework of financial markets which is also contained in the recently introduced realistic Order Book Model [T. Preis et al., Europhys. Lett. **75**, 510 (2006), T. Preis et al., Phys. Rev. E **76**, 016108 (2007)]. This model provides the possibility to generate the stylized facts of financial markets with a very limited set of rules. This model is described and analyzed in detail. Using this model, it is possible to obtain short-term anti-correlated price time series. Furthermore, simple profitability aspects of the market participants can be reproduced. A nontrivial Hurst exponent can be obtained based on the introduction of a market trend, which leads to an anti-persistent scaling behavior of price changes on short time scales, a persistent scaling behavior on medium time scales, and a diffusive regime on long time scales. A coupling of the order placement depth to the prevailing market trend, which is identified to be a key variable in the Order Book Model, is able to reproduce fat-tailed price change distributions.

1. Introduction

As a first-order approximation it is often assumed that price dynamics of financial markets obey random walk statistics. However, this rough approximation neglects the real nature of financial market time series, and a number of empirical deviations between financial market data sets and models consisting of random walk behavior have been observed [1][5]. From a physicist's point of view financial markets show a behavior just like a complex system in physics with non-Gaussian price increments. The price process of an exchange traded asset can be seen as the result of superimposed trading decisions of market participants. However, in context of modeling the "complex particle system" financial market with agent-based models using Monte Carlo techniques, market microstructure has a huge importance as well. "Market micro structure

[4] Supplementary information can be found on http://www.tobiaspreis.de

[5] In computer science applications and diverse interdisciplinary science fields such as computational physics [2] or quantitative finance[3], the computational power requirements led recently to general purpose computing on graphics processing units (GPU).

is the study of the process and outcomes of exchanging assets under explicit trading rules", as can be found in [4]. However, these trading rules changed in history. Thus, a historical outline is presented in the next section, in order to highlight the transformation from the beginning of floor trading to fully electronic market places in recent years.

1.1. Tulip mania in the United Provinces

A classic example for exchange based trading can be found in the famous tulip mania [5, 6, 7], which took place in the first part of the 17th century, in which tulips became a subject of speculation. This tulip mania had serious impact on the social life in the area which is known as the Netherlands today. The tulip mania is one of the first huge speculation bubbles in history. Tulips, which were originally only located in Asian countries, were introduced in Europe in the middle of the 16th century. In 1560, first tulips were transported from Istanbul, the court of the Ottoman Empire, to Vienna. There, botanist Carolus Clusius was responsible for the imperial botanical garden of Emperor Maximilian II, where he cultivated these plants in 1573 after importing 1,500 seeds. The tulip cultivation in the United Provinces, which are today known as the Netherlands, started then in 1593 after Carolus Clusius had come from Vienna and when he was able to breed tulips under the special, not ideal climate conditions in the Netherlands. This exotic flower started to enjoy great popularity and in the following time tulips became more and more the state of a status symbol. This fact triggered an avalanche and started a competition for possession of the rarest tulips. In the beginning the largest part of market participants at this "tulip market" can be found in the upper class of society. Tulips with lines and flames on the petals were particularly popular. However, this effect was often caused by a special virus, which infects only tulips and is known as the Tulip Breaking potyvirus.

This development can be seen as the starting point of a speculation bubble. In the following time, the competition for possession of tulips escalated and very high price levels were reached when the demands for tulip bulbs exceeded the offers. In 1623, a bulb of a very rare and preferred tulip could change hands for a thousand Dutch florins. This has to be seen in relation to the financial background — the average income per year was only about 150 Dutch florins. In the beginning, bulbs were only traded during the planting season. Later with increasing demand, bulbs could change hands during the whole year with the consequence, that bulbs were sold without the knowledge by the buyer of the exact tulip appearance. So, tulip trading became a speculation business. This aspect operated as a catalyzer for paintings in order to give impressions of the future appearance of a tulip to interested clients.

In the 1630s, the situation had come to a head. One reason was, that not only transactions with cash settlement were recorded. Tulips were also traded via option contracts. An option contract is a financial derivative and one can differentiate between call and put option. Buying a call option provides the right to buy a specified quantity of an underlying – in this case tulip bulbs – at a set price at some time on or before expiration of the contract, while buying a put option provides the right to sell the underlying at the specified price, the so-called strike price. These financial instruments make it possible to hedge price risks and allow producers to sell future crops. On the other hand, consumers can lock cheap buying prices for the future. As instrument of speculation, option contracts, which are traded at derivatives markets today, exhibit a high risk, as a large leverage effect is reachable with comparative small trading accounts. These Dutch option contracts were the first derivative instruments in history. And the increased use of option contracts in the Dutch tulip bulb trading can be regarded as a first evidence for a starting crisis. A high grade of debt of some market participants could be observed, which is one important ingredient for the formation of a speculation bubble. The prices of bulbs escalated with a factor of 50 in the years from 1634 till 1637 [7]. In Amsterdam, e.g., a mansion was sold for three tulip bulbs. And a transaction of 40 bulbs for 100,000 Dutch florins was observed and documented in 1635. A sale of a very rare tulip bulb – the Semper Augustus – was recorded with a price of 6,000 Dutch florins. These transactions have to be seen in comparison with the prevailing consumer price level in the United Provinces at that time. A ton of butter cost roughly 100 Dutch florins. This example clarifies the speculative nature of tulip bulb trading. In modern risk management, financial derivatives are reasonable instruments in order to control risks of a portfolio, which is a composition of various assets. In the Dutch tulip bulb crisis, however, option contracts were used in order to clear bottlenecks.

This bubble boosted the trading also in other parts of the Dutch society. In order to speculate in these upcoming markets, Dutch citizens of all ranks invested large parts of their wealth in tulips. As a result, some speculators made very large profits in that time and other people lost their whole wealth or even more. The speculation bubble burst in 1637. The inflated prices for tulip bulbs could no longer be realized by the tulip traders. The demand was falling which resulted in an upcoming panic. Everybody had the intention to sell, nobody bought tulip bulbs. The prices had reduplicated in January. And in February a top was reached. As in every bubble, there were also here winners and losers. Some people were holding open option contracts for selling tulips at prices, which were very high in relation to the current market value. The counterpart, who had the opposite position, had the duty to buy these bulbs for these utopian prices. Thousands of Dutch people were financially ruined.

1.2. History of exchange based trading

After this short outline of the first speculation bubble in history, one has to state, that the trading in the Dutch tulip mania was unorganized and a kind of a *over-the-counter* market (OTC)[6]. Contracts were concluded individually by the parties.

In the historical development of exchange based trading one can differentiate between spot markets and derivatives markets. Whereas spot market contracts, e.g. shares of a stock corporation, have to be fulfilled immediately[7], contracts on derivatives markets are fulfilled on a set date in the future, as already described in Section 1.1 for the option contracts. However, not only options can be traded at derivatives exchanges. Also an other standard group of contracts is available, future contracts. A future contract contains the right and the duty to buy or sell a specified quantity of an underlying at a set price at the expiration time of the contract.

First evidence of exchange based trading can be found in the tulip bulb bubble as mentioned before. At that time, options became important for economic processes for the first time. However, not only options on tulip bulbs were traded. Also contracts of the East India Company, which was founded in 1602, changed hands [8]. In Germany, options were dealable for the first time in 1970. However, in the United Kingdom and in the United States of America (USA), options were traded already in the 18th century. Despite of the fact, that the following outline is not exhaustive, the beginning of standardized futures trading should be mentioned, which took place at the Chicago Board of Trade (CBOT) in 1865. Additionally, despite of an established OTC market, the trading of exchange based stock options in the USA was possible at the Chicago Board Options Exchange (CBOE) since 1973. In Europe, the European Options Exchange (EOE) was founded in Amsterdam in the year 1978. And after teething troubles the EOE was established as the primary market for options in Europe in the early 1980s. This position was lost not until when the Deutsche Terminbörse (DTB) had been created in 1990 in Germany. In 1998, then the merger of DTB and Swiss Options and Futures Exchange (SOFFEX) led to a world leading derivative market [9], the European Exchange (EUREX), which was a pure electronic trading platform in contrast to established exchanges in the USA, which used a

[6] *Over-the-counter* markets (OTC) differ from exchange based trading, as they are not regulated. Furthermore, the fulfilling of OTC market contracts is not guaranteed by a central clearing house and so, the risk exists, that the contract counterpart is not able to fulfill an arranged contract.

[7] In the Federal Republic of Germany, spot market contracts will be cleared within at most two exchange trading days.

floor trading concept. There, traders exchange assets on the floor in a physical way. EUREX bought additionally the International Securities Exchange (ISE) in 2007, in order to play a pioneering task in the upcoming exchange consolidationprocess. However, this step has to be seen in context with power shifts in the large exchange companies all over the world. Also activities of hedge funds influenced this evolution, which were holding large stock packages of the Deutsche Börse AG[8] and the London Stock Exchange (LSE). However, the initial intention to force on a merger between both exchanges miscarried. Instead, a merger between the two spot markets New York Stock Exchange (NYSE) and Euronext was established and the NYSE Euronext was born – a large competitor of Deutsche Börse AG and LSE. The NYSE as a large spot market with a long tradition was founded in 1792. Up to March 2006, almost the whole trading on NYSE was handled on the floor. Then, with the acquisition of the electronic trading platform Archipelago, at NYSE the electronic era started. This process was enforced after fusion with Euronext. Today, a clear tendency to electronic market places with transparency can be observed. As could be read in September 2007 [10], the NYSE will close two more trading rooms on the floor. Simultaneously, one has to state that roughly one third of the trading volume at NYSE is generated by program trading[9] in the meantime. Program trading, which means the execution of automatic trading decisions generated by computer algorithms, can be realized on full-electronic trading platforms. Beside NYSE, XETRA – the electronic spot market platform of Deutsche Börse AG – should be mentioned, which replaced the old IBIS system in 1997 and started also a great success story.

A very impressive example for the change from floor trading to electronic trading can be found in trading of the Euro-Bund future[10] (FGBL), which is a fixed income derivative. In the beginning of the 1990s, the FGBL contract was traded at the young DTB and also at the 1982 founded London International Financial Futures Exchange (LIFFE), which is today part of NYSE Euronext. At that time, LIFFE was a pure floor trading exchange and the FGBL was favoredly traded at LIFFE. However, an empirical comparison between these two exchanges benefited the DTB, even though the transaction volume was lower than the one at LIFFE. One reason for the success of DTB was the higher level of transparency. The order book of the DTB, which stores offers and demands of all market participants and matches orders against each other, was distributed partly to the market members. Thus, each trader had an impression of the market depth. The advantage of anonymity boosted the electronic trading. With an increasing liquidity the DTB was able to concentrate the whole trading volume of the Euro-Bund future in its own trading system. This was one milestone for the triumphal procession of full-electronic exchange trading.

1.3. The Gold Fixing in London

However, even if electronic trading has mostly displaced floor trading and manual trading processes, some relics can still be observed today. Naturally, some exchanges indulge in floor trading for public relation reasons. The Deutsche Börse AG hangs on to floor trading. Recently, the accouterment was even modernized in order to present the general public a manifestation of financial markets. A daily trading process, which can be classified as a traditional institution, is the Gold Fixing in London [4]. The first Gold Fixing took place on September 12th, 1919 at 11.00 am in London in physical presence of the original five founding members N. M. Rothschild & Sons, Mocatta & Goldschmid, Samuel Montagu & Co., Pixley & Abell, and Sharps & Wilkins. The gold fixing is executed twice a day. At that time, the five members came together. Since

[8] EUREX is a joint venture of Deutsche Börse AG and Swiss Exchange (SWX).

[9] Detailed information can be found in the news releases of NYSE Euronext on the website `www.nyse.com`.

[10] The underlying of the Euro-Bund future contract is a certain debt security issued by the Federal Republic of Germany with a remaining term of 8.5 to 10.5 years.

2004, a dedicated conference call is used[11]. In the beginning of each fixing, the Chairman of the meeting, who is one of the members, announces an opening price. The other four members have then the possibility to contact their customers and include their orders into their decision. Then, each member has to declare if the member institution is a buyer or a seller at the announced price level. In the case, that there are buyers and sellers, additionally the quantity offered and demanded at this price is asked in units of bars. If only buyers or only seller exist at the announced price, or if the difference between offered and demanded bars at a price level is larger then 50 bars, then the procedure starts again and the price is moved. If a price finally fulfills the requirements, the chairman announces that the price is fixed. In order to have the possibility to contact clients, each member has a verbal flag. As long as a flag is raised, a price cannot be declared fixed. However, this verbal flag has physical roots. Before the dedicated conference call was established, each member had a small Union Flag on the desk for this purpose. Since 1919, the gold price fixing provided market participants with the opportunity to buy and sell gold at a specific price. The London Gold Fixing is still today a benchmark for trading with gold. However, as seen above, former procedures were adapted to progress in information technology. One has to state, that there exists full-electronic trading of gold today. On the New York Mercantile Exchange (NYMEX) founded in 1882 and on the New York Commodities Exchange (COMEX), which is a sub division of NYMEX today, gold can be traded continuously.

1.4. Econophysics – Physics meets Finance

In the remainder of this Section, in a short outline, the milestones of modeling financial markets should be presented, especially in context of physicists' contributions. Some decades ago, M. F. M. Osborne [11, 12] and B. B. Mandelbrot [13, 14] started to apply methods and also models from statistical physics to problems arising in the examination of the behavior of financial markets, and thus they became founders of a new research field, which is nowadays often called econophysics. They discovered that the price fluctuations at financial markets do not obey random walk statistics [15, 16, 17], as assumed in many economic theories. However, the modeling of financial market time series by a diffusive stochastic process dates already back to L. Bachelier [18] in 1900. Geometric Brownian motion [19, 20, 11], which is a variant of the Brownian motion and which was independently investigated by L. Bachelier and A. Einstein [21], has become the standard mathematical model, and the theory of fair prices of option contracts [22, 23] is based on it. With this theory, it was possible to find a fair value for options for the first time – a huge progress since option trading in the Dutch tulip bulb crisis. However, also a sophisticated model does not protect against losses. Two originators of the theory, namely Black and Merton, founded the hedge fund Long Term Capital Management (LTCM) in 1994 in order to exploit advantage derived from the theory. However, after an initially enormous success, the fund lost more than 4 billion dollars in 1998 during the Russian Financial crisis. It is an often found fact, that in extreme events, models lose their stability and sometimes their validation.

The insights found by B. B. Mandelbrot were based on time series of very limited length. With the progress in information technology, trading processes at the international exchanges were adapted to the available IT infrastructure and full electronic exchanges like EUREX were created. The trading at full electronic exchanges can produce time series data down to centiseconds base and thus nowadays a huge quantity of historical financial market time series is available. By using these new possibilities, the work of B. B. Mandelbrot could be confirmed. Noticing discrepancies between economic theory and financial market reality, physicists started approaching the complex "multi particle systems" of financial markets and economy in general with a large variety of physical methods. Recent examples can be found in [24, 25]. An early contribution to this enterprise was provided by P. Bak [26]. His very simple agent based model

[11] More information can be found on the website `http://www.goldfixing.com`.

uses the reaction diffusion process $A + B \rightarrow 0$ and contains elements of trader imitation and feedback mechanisms. Nowadays, physicists are interested in understanding the process of price formation in more detail, as former models were not able to reproduce the behavior of financial markets completely. In this context, the model of S. Maslov [27, 28] has to be mentioned. He published an alternative model, in which two different types of orders were introduced. Further agent based market models [29, 30, 31, 32, 33, 34, 35, 36] were developed with the aim to reproduce and interpret empirical stylized facts of financial market time series. Especially a model of J. D. Farmer et al. [37, 38] has to be mentioned. In this statistical model, the continuous double auction is examined analytically and numerically.

In the following Sections, the Order Book Model, which was introduced in [39] and analyzed in detail in [40], is presented in combination with the introduction of the order book structure. This model can be used in order to obtain a "mechanistic understanding" of the price formation process. Some extensions are given leading to empirical stylized facts, which can be observed on real financial markets.

2. Market Structure and Order Book Model

As described in the introduction, electronic exchanges are the prevailing form of trading nowadays. Upon this background, the Order Book Model [39, 40], which is based on structures found at real electronic financial markets, is defined in its basic form in this Section. Aim is the accurate reproduction of structure and the mechanisms of an order book at real financial markets, as shown in Fig. 1. The function of an order book is to store offers and demands of the market participants. In the simulations, only one order book is used, in which one individual asset is traded. This asset can be, e.g., a share of a stock company, a loan, or a derivative product. There are also various types of orders at real financial markets. In the Order Book Model, only the two most important types are implemented, namely limit orders and market orders. Limit orders are executed only at a specific price or a price, which is better for the trader, who placed this limit order, whereas market orders are executed immediately against the best available limit order. Limit orders are added chronologically, which realizes a price time priority based matching algorithm. Thus, at a given price level with more than one limit order, the limit order, which was inserted first, has execution priority. Limit and market orders are performed by agents in the Order Book Model, which also distinguishes between liquidity providers and liquidity takers. These two groups of agents differ in the types of orders they are permitted to submit. On the one hand, N_A liquidity providers transmit only limit orders. In the case of a limit sell order, a liquidity provider offers an asset for sale at a set limit price or a higher price. Analogously, a limit buy order indicates a demand for buying an asset and the order is executed at a set limit price or any better price for the liquidity provider. Let p_a be the so-called best ask price, which is the lowest price level of all limit sell prices in the order book, and analogously p_b the so-called best bid price, being the highest price level for which at least one limit buy order is stored in the order book. In the Order Book Model, limit orders are placed around the midpoint $p_m = \frac{p_a+p_b}{2}$ with a rate α, i.e., $\alpha \cdot N_A$ new limit orders are submitted per time step. Let q_{provider} be the probability with which a limit order is a limit buy order. Thus, with probability $1 - q_{\text{provider}}$, the limit order to be placed is a limit sell order. The liquidity provider, which can be identified as so-called market maker, supplies liquidity to the market in order to exploit the non-zero spread $s = p_a - p_b$ for earning money: these market participants intend, e.g., to sell an asset at price p_a or higher and then to buy it back at price p_b or lower. Thus, they have earned at least the spread s. As seen in this example, so-called short sales are allowed, i.e., it is allowed for agents to sell assets even if they do not possess them.

On the other hand, there are N_A liquidity takers, who transmit only market orders. These market orders are submitted with rate μ, i.e., $\mu \cdot N_A$ market orders are inserted per time step into the order book. A market order will be immediately executed after arrival. A market sell order

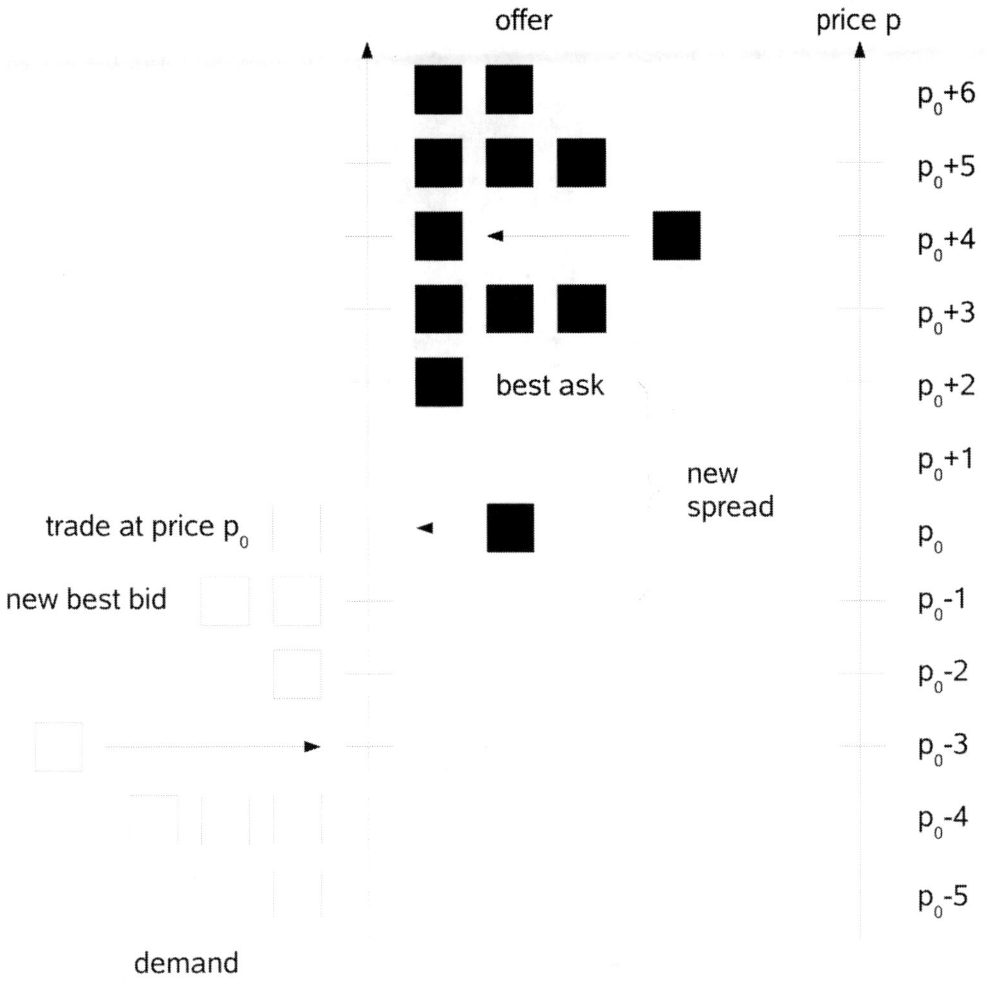

Figure 1. Market micro structure of electronic order books: Limit buy (white squares) and limit sell orders (black squares) are added chronologically to the appropriate discrete price level p. In this example, a new sell order is placed at price p_0, at which there is already a buy order stored in the order book. This limit sell order, which can be executed immediately against the buy order on the bid side, is a so-called crossing limit order and so the order degenerates to a market order. With the execution of the two orders at price p_0, a trade is performed at price p_0, which is then called last traded price. The spread s, which is the difference between the best ask and the best bid price, was two ticks before the arrival of the crossing limit sell order at price level p_0 and increases to three ticks after this trade, as shown in Fig. 2.

is executed at price level p_b, a market buy order at price level p_a. A market order of a liquidity taker is a market buy order with probability q_{taker} and a market sell order with probability $1 - q_{\text{taker}}$. In the basic version of the Order Book Model, the simple case $q_{\text{provider}} = q_{\text{taker}} = \frac{1}{2}$ is applied. Thus, all orders will be produced symmetrically around the midpoint. In practice, it is possible, that the limit price of a limit sell order is lower than the current best bid price and the limit price of a limit buy order is higher than the current best ask price. Such so-called crossing

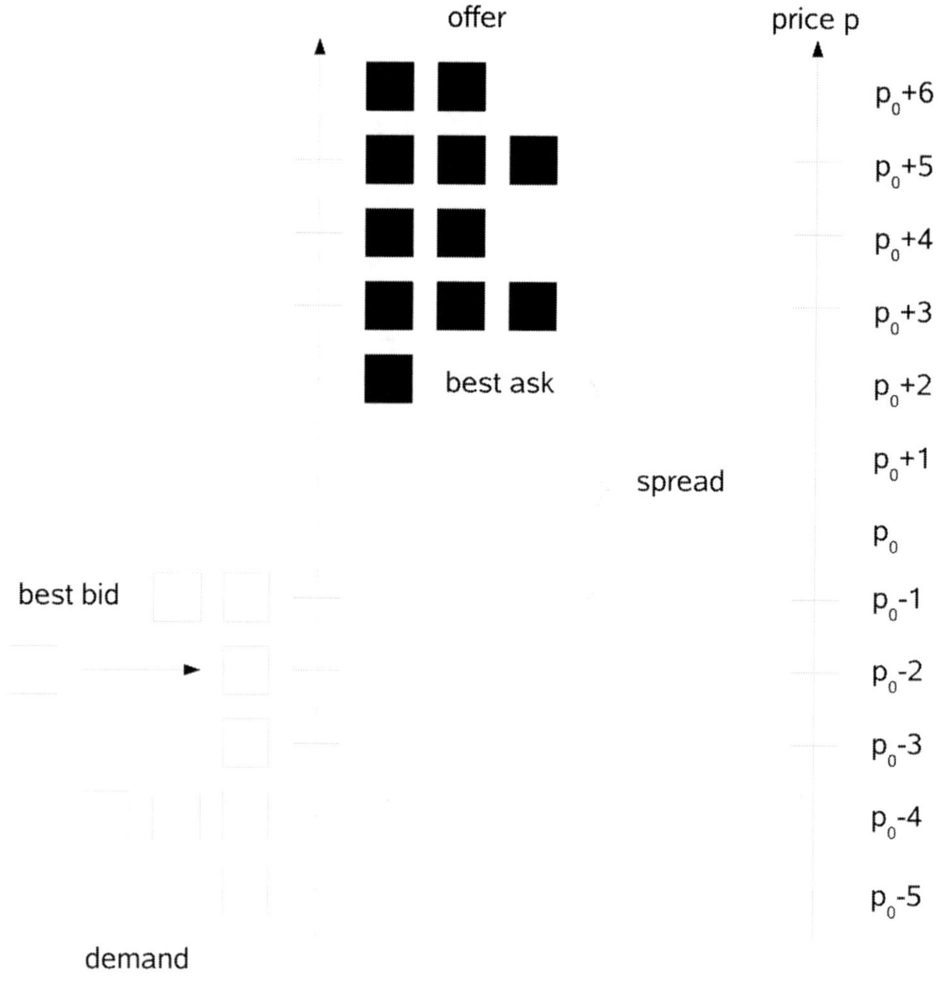

Figure 2. Market microstructure of electronic order books: This figure shows the order book after the execution in Fig. 1. The new spread is three ticks.

limit orders degenerate to market orders and thus, they are executed immediately. In the Order Book Model, only pure limit orders will be used. Limit orders, which are stored in the order book, can also expire or can be deleted. In the model, this canceling is realized in the way that each stored order is deleted with probability δ per time unit. As there are overall $2N_A$ agents in the multi-agent system, each Monte Carlo step (MCS) consists of $2N_A$ moves. In each move, one agent is randomly selected and can perform one action according to the probability rates. If the chosen agent is a liquidity provider, then a limit order with probability α is submitted by this agent. On the other hand, if the selected agent is a liquidity taker, then a market order with probability μ is placed in the order book and will be immediately executed. Orders in the Order Book Model have the constant order volume 1. Thus, it is possible only to buy or sell one asset unit, e.g. one share, with a single order.

Based on these simple rules, first an unrealistic independent identically distributed (IID) order placement depth is applied. This is realized in the way that limit buy orders are entered on each price level in the interval of $[p_a - 1 - p_{\text{int}}; p_a - 1]$ with the same probability, and accordingly, limit sell orders are transmitted uniformly distributed in the interval of $[p_b + 1; p_b + 1 + p_{\text{int}}]$. Already with this definition of the Order Book Model, profits and losses of the agents can be analyzed. And one can find, that the averaged wealth value of liquidity takers and liquidity providers drifts apart linearly in time [40]. If comparing these results with real financial markets, it has to be stated that liquidity takers are disadvantaged compared to liquidity providers systematically. The distinction in the Order Book Model between the two groups of market participants reflects the two types of orders. In real financial markets, each market participant is not restricted to one order type.

In the next step, a more realistic order placement depth will be integrated in the Order Book Model. The order book depth of real financial markets can be described by a log-normal distribution [28]. And, to take this into account the IID limit order placement is replaced by an exponentially distributed order placement depth. Thus, for placing a limit order, the limit price p_l is determined for a limit buy order through

$$p_l = p_a - 1 - \eta \tag{1}$$

and for a limit sell order according to

$$p_l = p_b + 1 + \eta \tag{2}$$

whereby η is an exponentially distributed integer random number created by $\eta = \lfloor -\lambda_0 \cdot \ln(x) \rfloor$ with x being a uniformly distributed random number in the interval $[0; 1)$ and $\lfloor z \rfloor$ denoting the integer part of z. With this construction, the submission process of limit orders has the tendency to reduce the gap between best bid price and best ask price. Also, crossing limit orders are avoided, as the price of a limit buy order cannot be equal or larger than the best ask price and the price of a limit sell order cannot be equal or lower than the best bid price.

As result of applying the exponential order placement rule, a log-normally distributed order book depth profile is obtained, as shown in Fig. 3. As order book depth profile, one describes the cumulative order volumes of the orders at each price level of the order book. In Fig. 4, an exemplary price sequence of 10^6 MCS is shown. The price history of an Order Book Model simulation is used in order to analyze the most important properties of price time series.

One important time series characteristic can be measured by the Hurst exponent, which is shown in Fig. 5 for various agent numbers N_A. Generally the Hurst exponent $H(q)$ can be calculated by the relationship

$$\langle |p(t + \Delta\tau) - p(t)|^q \rangle^{1/q} \propto \Delta\tau^{H(q)}, \tag{3}$$

as defined for example in [41]. If it is not mentioned otherwise $H(\Delta\tau) \equiv H(\Delta\tau, q = 2)$ is used. Mainly, the Hurst exponent provides a measurement of the scaling behavior of a process. $H < 0.5$ indicates an antipersistent behavior, i.e., that after a movement, a reversion tendency can be found. A Hurst exponent larger than 0.5 can be found for persistent processes. There the probability for continuation of a movement is larger than for a reversion. The random walk process exhibits $H = 1/2$. For comparison, the Hurst exponent of the random walk, which is constantly $1/2$ for all time lags, is shown in Fig. 5, too. On short time scales, the price process provides an antipersistent behavior, which is a consequence of the order book structure. On long time scales, the price process converges towards random walk properties. The antipersistence on short time scales, which can also be found in real financial market data, is a consequence of the short time anticorrelation. The autocorrelation of the dedicated price changes time series $\delta p(t) = p(t + 1) - p(t)$ shows the same behavior as can be found in real financial markets. A negative autocorrelation exists for the time lag $\delta\tau = 1$. Thus, a positive price change succeeds

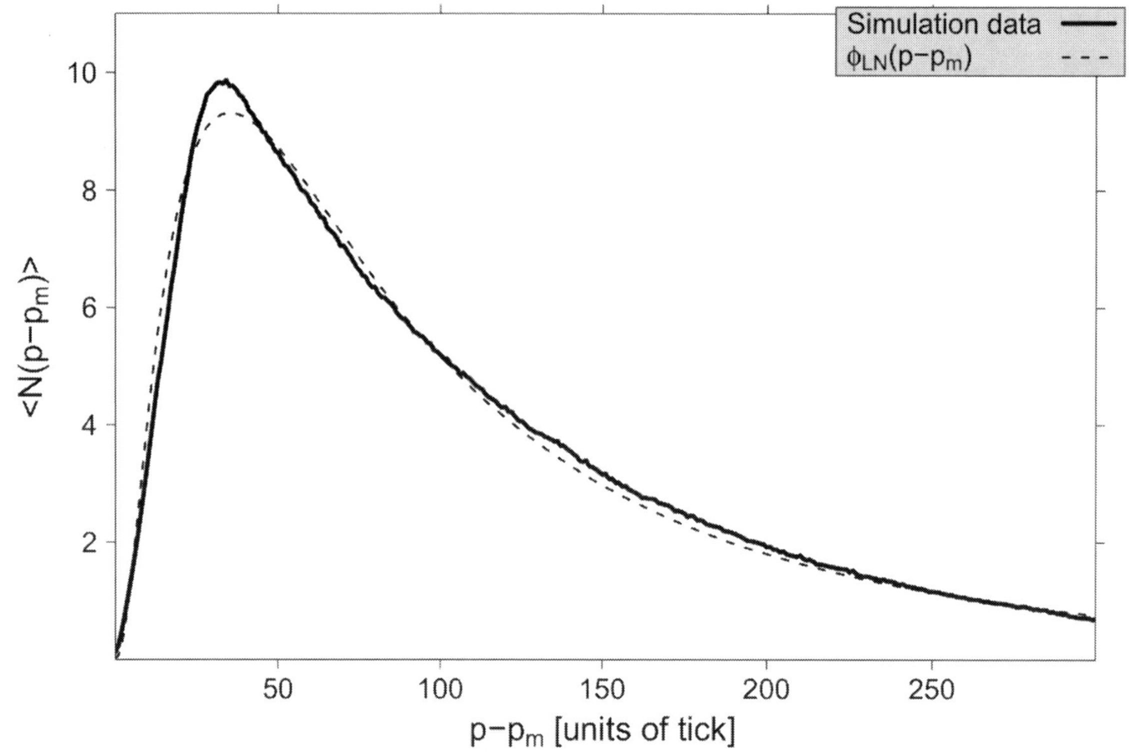

Figure 3. For an exponential order placement depth and with the parameter selections $\alpha = 0.15$, $\mu = 0.025$, $\delta = 0.025$, $\lambda_0 = 100$, $N_A = 250$, and $q_{\text{provider}} = q_{\text{taker}} = 0.5$, one obtains the shown order book depth. The results were averaged over 10^4 MCS. A fit with the log-normal distribution function $\phi_{\text{LN}}(p-p_m) = a/((p-p_m)\cdot\exp\{-[\ln(p-p_m)-b]^2/c\}$ leads to the parameters $a = 527 \pm 1$, $b = 4.483 \pm 0.004$, and $c = 1.79 \pm 0.01$.

a negative price change with large probability and vice versa. This is a consequence of the "mechanics" of the order book. As there are constant order rates and equal probabilities for buy and sell orders, an executed market order of any kind increases the probability that the next last traded price will be anti-correlated with the previous price. An other most important stylized fact of financial market time series is the property, that price change distributions show non-Gaussian shapes. In fact, the probability for large price changes in any direction is larger than predicted by the Gaussian distribution. The price change distributions for this basic implementation of the Order Book Model are shown in Fig. 6. These distributions exhibit no fat tails. However, they can rather well be approximated by a Gaussian distribution.

3. Parameter space

In the previous section, results for a specific parameter selection were presented. However, in the Order Book Model, a large number of essential control parameters can be found. And in order to apply the Order Book Model to a large variety of situations, it is necessary to approach the parameter space of the Order Book Model. The important order rates – the limit order rate α and the market order rate μ – and also the characteristic order entry depth λ_0 can be tuned. It could be found that the order book depth profile is of crucial importance. If the density of order book depth is too high, then the price process has the tendency to freezing. If the density is too low, then randomly consecutive market orders of the same kind, i.e. either only market buy orders or only market sell orders, can empty the contrary side of the order book, which

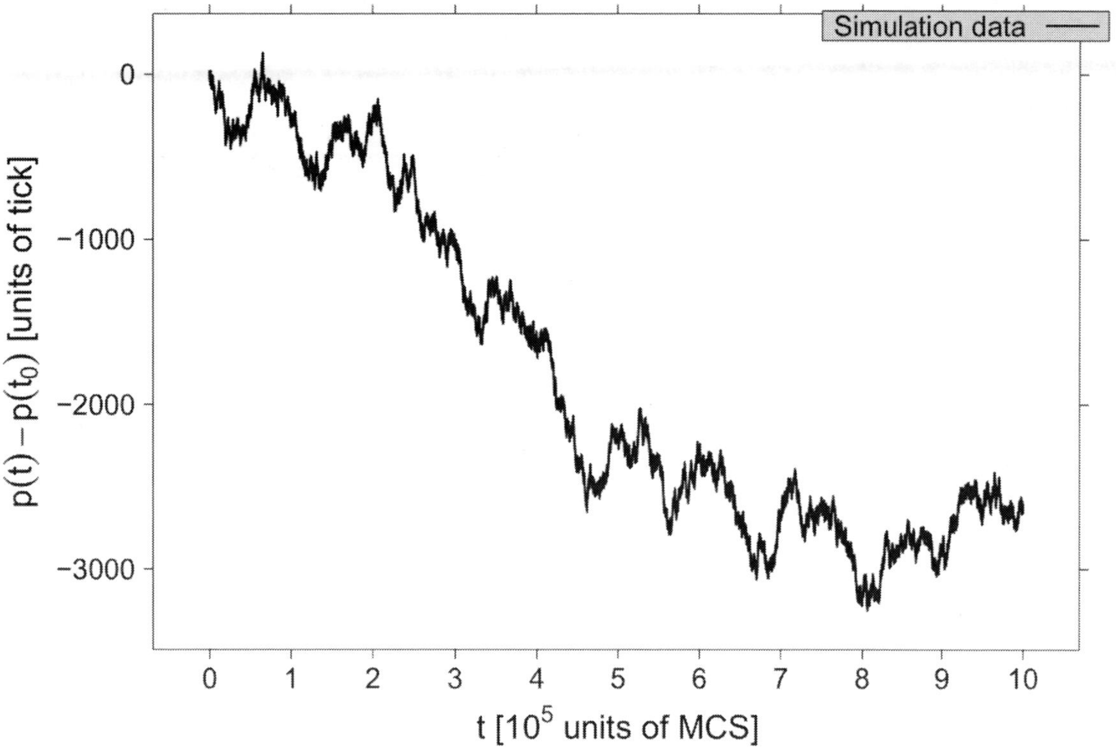

Figure 4. For an exponential order placement depth and for the parameter selection $\alpha = 0.15$, $\mu = 0.025$, $\delta = 0.025$, $\lambda_0 = 100$, $N_A = 125$, and $q_{\text{provider}} = q_{\text{taker}} = 0.5$, an exemplary price sequence of 10^6 MCS is shown.

was previously filled with limit orders. For this reason, the order book depth profile is analyzed quantitatively in dependence of the two most important parameters λ_0 and μ with fixed α, δ, and N_A. After a short transient time of few thousands MCS, the order book depth is stationary and log-normally distributed according to

$$P_{\text{LN}}(x) = A \frac{1}{Sx\sqrt{2\pi}} \exp\left(-\frac{(\ln x - M)^2}{2S^2}\right) \tag{4}$$

with the parameters A, S^2, and M. In Fig. 7, one can find the relation between market order rate μ and scaling factor A of the log-normal distribution for various values of λ_0. However, this prefactor A can be theoretically obtained, too. A determines the integral over the log-normal distribution function and thus corresponds to the total number of limit orders stored in the order book. The number of orders in the order book at time step $t+1$ can be calculated iteratively by the various order rates α, δ, μ, the number of orders $N(t)$ at the previous time step t, and by the number of agents N_A acting in the system, through

$$N(t+1) = N(t) + \alpha N_A - (N(t) + \alpha N_A)\,\delta - \mu N_A. \tag{5}$$

In equilibrium state,

$$\frac{N_{\text{eq}}}{N_A} = \alpha\left(\frac{1}{\delta} - 1\right) - \frac{\mu}{\delta} \tag{6}$$

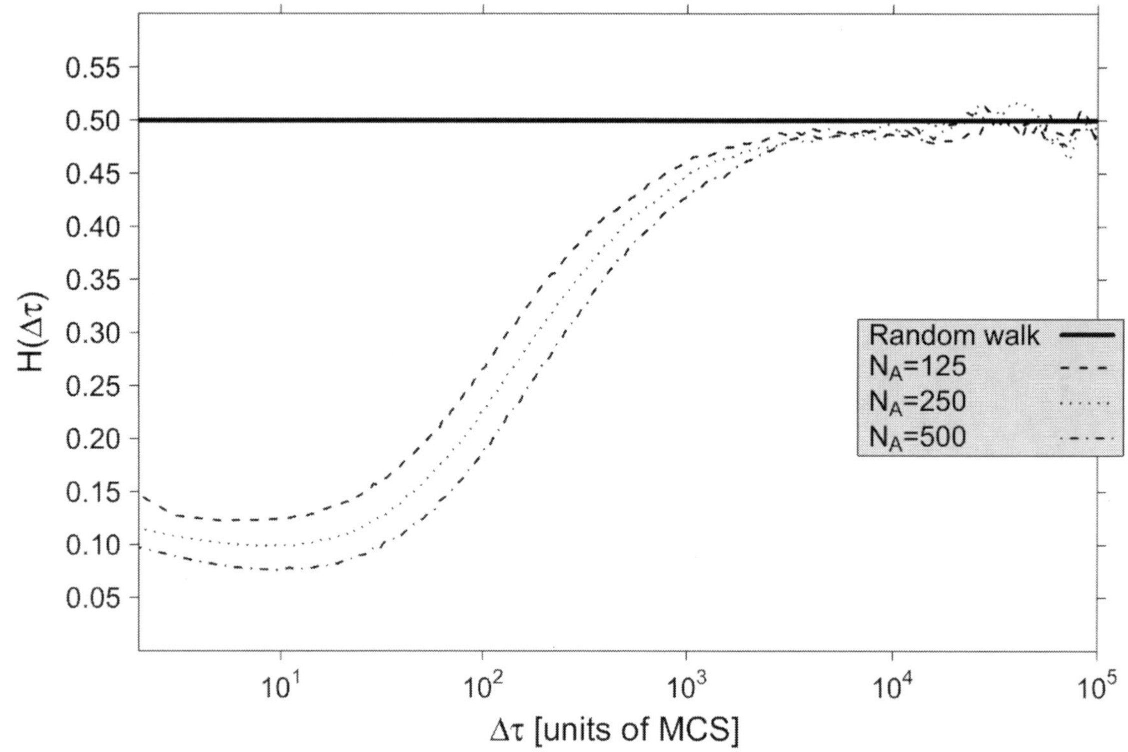

Figure 5. For the same parameter selections as in Fig. 4, the Hurst exponent is shown for various numbers of agents N_A, averaged over 50 simulation runs each.

can be obtained for the number of stored orders per liquidity provider. One can additionally define an effective limit order rate $\alpha^* = \alpha\,(1-\delta)$. Then, the expression

$$\frac{N_{\rm eq}}{N_A}\,\delta = \alpha^* - \mu \tag{7}$$

provides, that a stable order book is only then achieved, if the conditions $\alpha^* > \mu$ and $\delta > 0$ are fulfilled. When fitting the order book depth to the log-normal distribution, only one half of the orders is considered, leading to an expression for $N_{\rm eq}/2$ of $1462.5 - 10^4\mu$, calculated by the values used in the order book simulations. This expression is given additionally in Fig. 7 as a solid line. It can be found that the theoretical considerations fit to the simulation results on medium and large values of the market order rate μ. A linear decrease of A for medium and large values of μ has to be stated. And this behavior is independent of the parameter λ_0. The deviation between theory and simulation found for small μ results from the fact that the order book depth does not approach a log-normal distribution for very small market order rates μ. Then, a fit to the log-normal distribution function is of course no longer valid.

A second important parameter, which is influenced by changes of the parameters λ_0 and μ, is the center point M of the log-normal distribution. The larger the market order rate μ, the more limit orders can be removed from the bid side and the ask side of the order book by order matching. The more limit orders, however, are removed from the inner part of the log-normal distributions at both sides of the order book, the more the center point M is departing from the midpoint p_m being between the two distributions. This is illustrated in Fig. 8. This phenomenon can be found for $\mu > 0.01$. Deviations from the log-normal behavior are considered, e.g., for small values of μ. In this case, the distribution of the order book depth resembles the exponentially

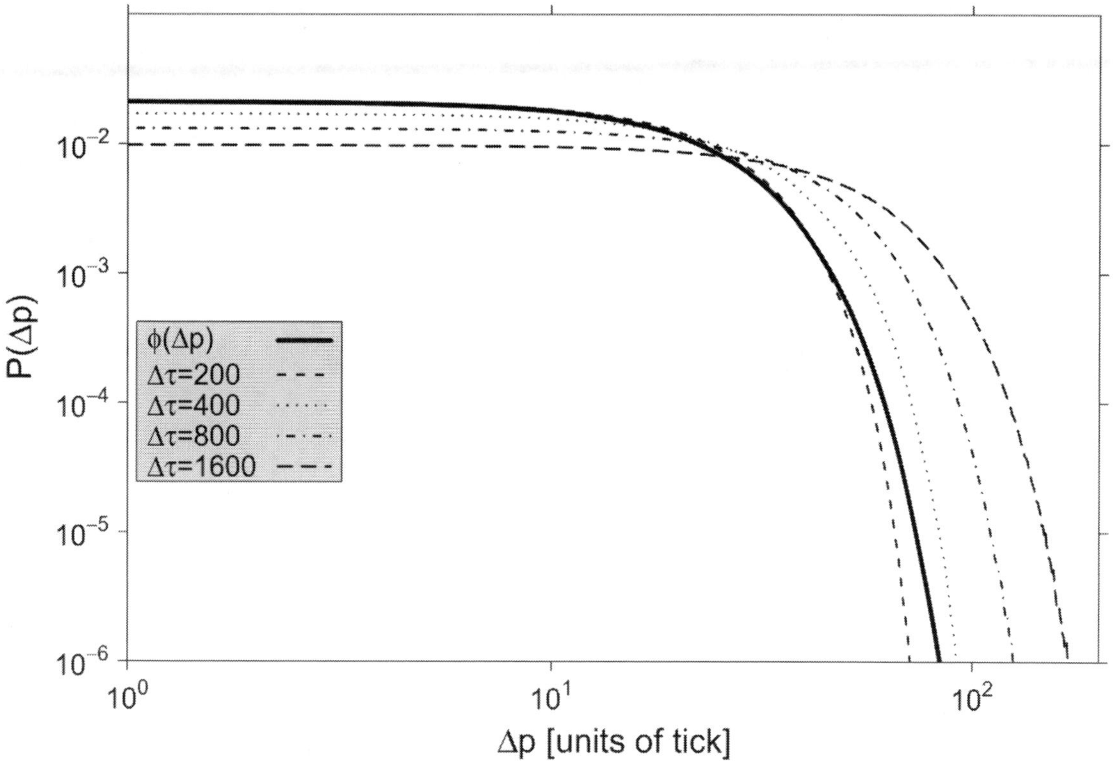

Figure 6. Distributions of price changes for various $\Delta\tau$ for the same parameter values as in Fig. 4. Again, the results were averaged over 50 simulation runs each. For $\Delta\tau = 200$, the fit parameters for the Gaussian distribution function $\phi(\Delta p) = a \cdot \exp(-b \cdot \Delta p^2)$ are given by $a = 2.122 \times 10^{-2} \pm 4 \times 10^{-5}$ and $b = 1.388 \times 10^{-3} \pm 10^{-6}$.

distributed order placement depth used for submitting the limit order. The reason for this observation is that the number of market orders is not sufficient in order to remove the limit orders at best bid price and best ask price. In the extreme case of vanishing μ, the order book freezes. Best bid price and best ask price become constant in time and the price jumps between best bid price and best ask price, if there are still market orders. Another extreme situation occurs by large market order rates μ. Then, the increased number of submitted market orders increases the probability that one side of the order book is completely cleared and then trading will stop.

4. Augmentation of the Order Book Model

The previous version of the Order Book Model is able to reproduce the short time antipersistence and the random walk behavior on long time scales. After the investigation of the parameter space, it is obvious, that some extensions of the basic version of the Order Book Model are required in order to reproduce further stylized facts of financial markets. So far, a persistent behavior on medium time scales cannot be reproduced. In the basic version of the model, always a symmetry between the buy and the sell probability is applied by $q_{\text{provider}} = q_{\text{taker}} \equiv 1/2$. This setting is in accordance with the assumption that financial markets fulfill the properties of a stationary behavior. However, a stationary state cannot be found at real financial markets. In fact, asymmetries are indicated by variable order rates, also on intra-day time scales. In short time trend phases, an increased buy or increased sell probability causes the movement of price

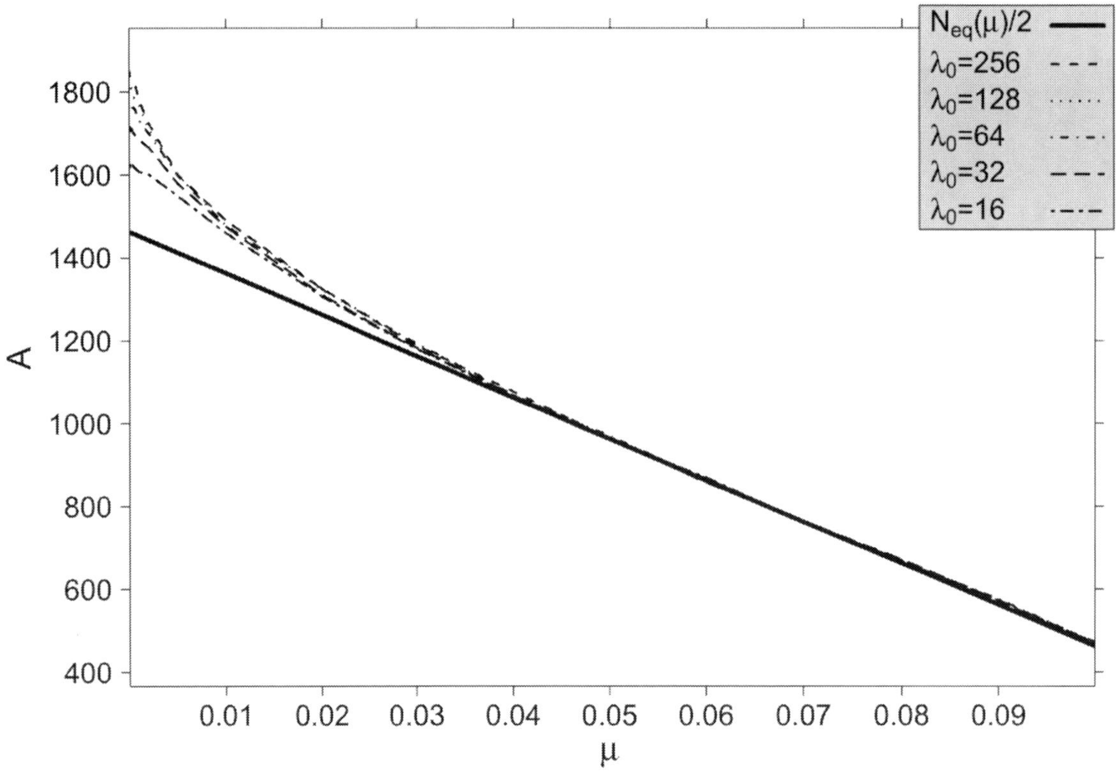

Figure 7. Fit parameter A as a function of the market order rate μ for various values of the order placement depth λ_0.

in one direction. Therefore, the symmetry $q_{\text{provider}} = 1 - q_{\text{provider}} = q_{\text{taker}} = 1 - q_{\text{taker}} = 1/2$ will now be broken in a way that q_{taker} shall be changed in time but still have an average value of $1/2$, in order to protect the order book stability. With $q_{\text{provider}} = 1/2$, the buy and sell probability of liquidity providers stays in equilibrium. With this method, the market in the simulation of the Order Book Model is "deflected" from its stationary state. One possible ansatz for the realization of such a market deflection is a deterministic symmetry disturbance. However, it can be shown [40], that the deterministic perturbation is not able to produce a more realistic price behavior. But it can be stated, that it is possible to change the behavior of the Hurst exponent especially on medium time scales. As deterministic perturbations with a discrete spectrum for the returning time of q_{taker} to the start value $1/2$ lead to oscillations of the Hurst exponent, a stochastic perturbation with continuous return time distributions is introduced in the next section.

4.1. Feedback random walk

A first stochastic perturbation is realized by a bounded random walk approach [39]. For this ansatz, the value of q_{taker} is changed in time by a bounded random walk with reflecting boundary conditions and with an average value of $1/2$. The buy and sell probability of liquidity providers stays constant through $q_{\text{provider}} = 1/2$. This extension of the Order Book Model leads to an antipersistent price behavior on short time scales, a persistent price behavior on medium time scales, and a random walk behavior on long time scales. One critical aspect is, that the maximum value of $H\,(\Delta\tau)$ is too large. As described in [13, 14, 42], only maximum values of approximately 0.6 should be expected. However, the shape of the price change distributions is dramatically

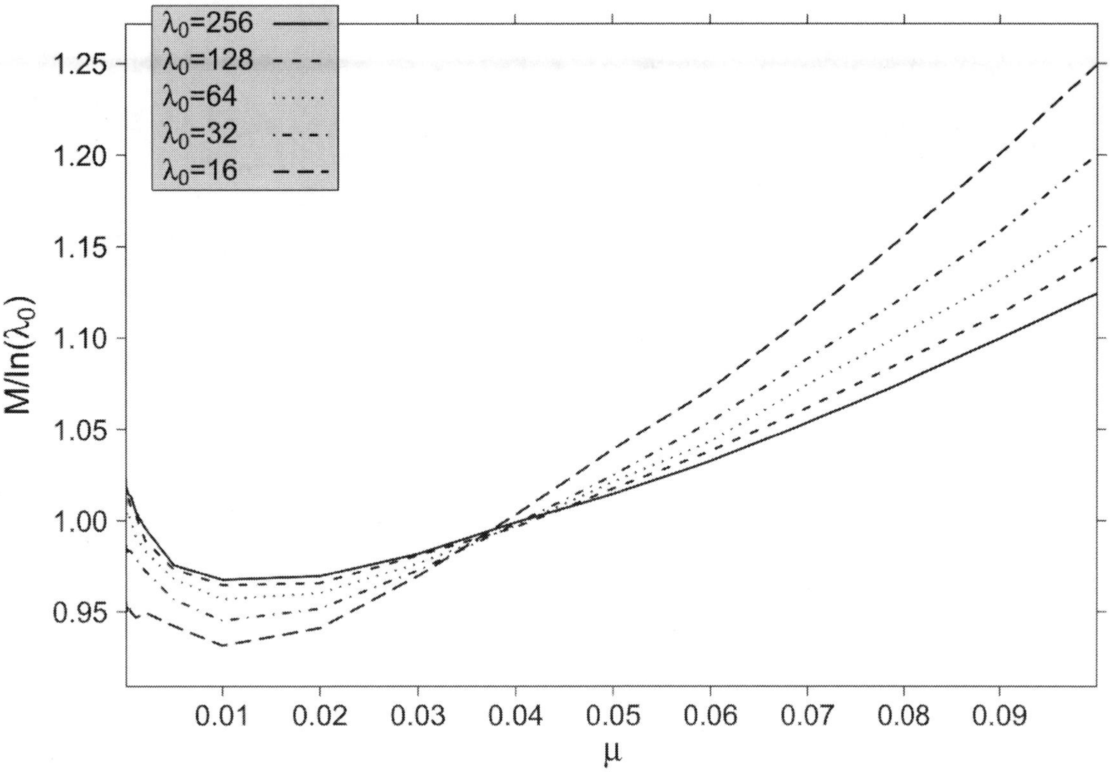

Figure 8. Fit parameter M as a function of the market order rate μ for various values of the order placement depth λ_0.

changed. A bimodal shape can be observed, which is in conflict with real financial markets. This behavior is caused by the constant residence distribution of a bounded random walk [40]. For this background, a sophisticated solution is a feedback random walk (FRW) – a random walk with increased probability for returning to the mean value. The FRW is again only applied to q_{taker}; q_{provider} stays constant at $1/2$. In the following, the FRW process is described in detail. In the beginning, q_{taker} starts at the mean value of $1/2$. Then, with proceeding time, q_{taker} is incremented and decremented by a value of ΔS in each Monte Carlo step. The probability for drifting to the start and average value of $1/2$ is given by $1/2 + |q_{\text{taker}}(t) - 1/2|$. Thus, the probability for departing from the start and mean value of $1/2$ is given by $1/2 - |q_{\text{taker}}(t) - 1/2|$. With these rules, the FRW tends to return to its start and average value $\langle q_{\text{taker}} \rangle = 1/2$. This FRW process obeys a continuous return time spectrum, which is qualitatively comparable to that of the bounded random walk. However, the residence distribution of the probability q_{taker} exhibits an almost Gaussian shape for the FRW, whereas the bounded random walk exhibits uniformly distributed residence.

If applying the FRW mechanism in the Order Book Model simulations, one obtains the behavior of the Hurst exponent as shown in Fig. 9. The Hurst exponent $H(\Delta\tau)$ was averaged over 50 simulation runs each consisting of 10^6 Monte Carlo steps (MCS). As already found in the basic version of the Order Book Model an antipersistent behavior on short time scales and a diffusive regime on long time scales is obtained. Additionally a persistent regime on medium time scales can be found. The maximum value of the Hurst exponent increases with increasing agent numbers N_A. In Fig. 10, the price change distributions for the FRW approach are presented, which are almost Gaussian shaped. Thus, also this extension of the Order Book

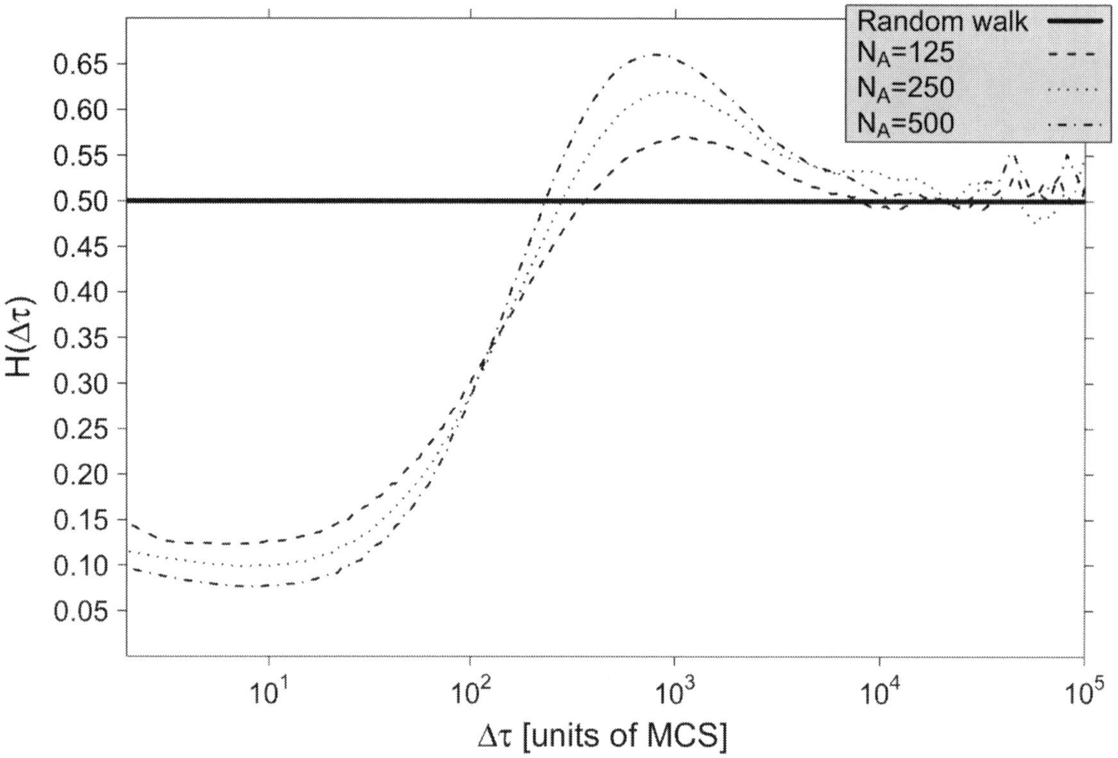

Figure 9. Feedback random walk perturbation of q_{taker}: Hurst exponent $H(\Delta\tau)$ for various agent numbers N_A as a function of $\Delta\tau$. The parameters used are $\alpha = 0.15$, $\mu = 0.025$, $\delta = 0.025$, $\lambda_0 = 100$, and $\Delta S = 1/1000$.

Model fails in producing fat tailed price change distributions. However, the fact, that $H > 0.5$ is obtained on medium time scales, can be interpreted in the context of autocorrelations. The Hurst exponent is often mentioned in order to characterize the behavior of time series and stochastic processes. Oftentimes a connection to the autocorrelation property is drawn. In literature, it is widely assumed, that $H \neq 1/2$ implies long time correlations. However, it was recently shown theoretically in [43] that $H \neq 1/2$ implies not necessarily long time correlations. The Order Book Model supports this statement, as a persistent regime can be found on medium time scales in combination with no non-vanishing autocorrelations on these time scales [40]. The autocorrelation function of price changes indicates a negative autocorrelation for time lag $\Delta\tau = 1$ in agreement with the antipersistent behavior on short time scales [39]. The autocorrelation functions for the quadratic price changes and for the absolute price changes are positive on short time scales and converge roughly exponentially towards zero [40]. This is in agreement with the so-called volatility clustering found in real financial market data as analyzed in detail in [44, 45, 46]. A posteriori the volatility can be simply calculated by the standard deviation of the values in the price time series.

4.2. Dynamic order placement depth

In order to motivate a further extension of the Order Book Model, one has to discuss risks existing for the market participants. A liquidity provider has the intention to earn money by placing a limit buy order at or near the best bid price and by placing a limit sell order at or near the best ask price. Under beneficial conditions, both orders are executed in a short time-frame.

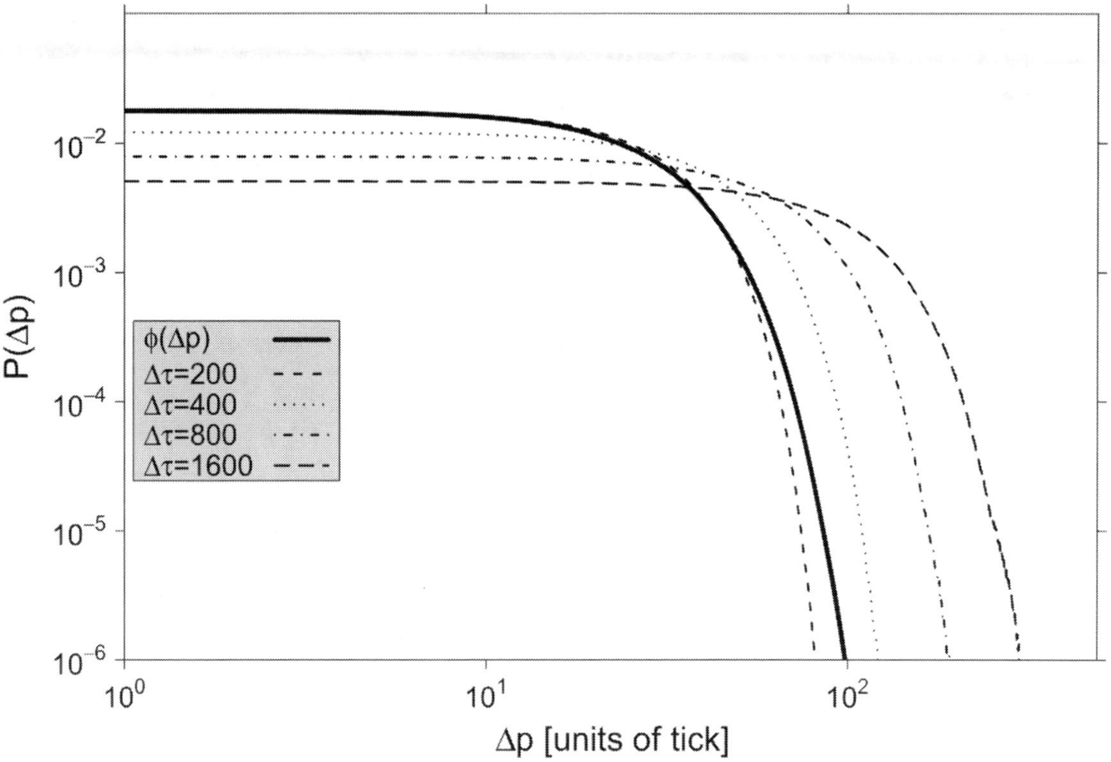

Figure 10. Feedback random walk perturbation of q_{taker}: price change distributions for various values of $\Delta\tau$ and for $N_A = 500$. The other parameters are identical as in Fig. 9. For $\Delta\tau = 200$, the function $\phi(\Delta p) = a \cdot \exp(-b \cdot \Delta p^2)$ is fitted to the data. Then, the fit parameters $a = 1.79 \times 10^{-2} \pm 10^{-4}$ and $b = 9.9 \times 10^{-4} \pm 10^{-5}$ are achieved.

However, it is also possible, that only one order, e.g., the limit sell order is matched against a market buy order. If after the execution, a trend with increasing prices is established, then the liquidity provider has only a small probability to close his position by getting an execution for the limit buy order stored in the order book. Instead, the liquidity provider is losing money during the time in which the market is rising. Concerning this aspect it follows, that liquidity providers can reduce their risk by adapting their limit order placement to the prevailing market conditions. In trendless market phases, in which only small price fluctuations are to be expected, liquidity providers can place their limit orders close to the midpoint, in order to be able to participate in these movements. However, in strong trend phases, the risk of the liquidity providers to possess positions, which are orientated against the prevailing market trend increases. Therefore, it is a consequence, that a liquidity provider adapts the characteristic order placement depth in order to take into account changing conditions. In the Order Book Model, the strength of a trend is related to the deviation of the variable q_{taker} from the symmetric case $q_{\text{taker}} = 1/2$. Therefore, the constant order placement depth λ_0 is replaced by:

$$\lambda(t) = \lambda_0 \left(1 + \frac{\left| q_{\text{taker}}(t) - \frac{1}{2} \right|}{\sqrt{\left\langle \left(q_{\text{taker}}(t) - \frac{1}{2} \right)^2 \right\rangle}} \cdot C_\lambda \right) \tag{8}$$

This extension of the Order Book model corresponds to $C_\lambda = 0$ in the original version of the Order Book Model. In Fig. 11 and Fig, 12, the Hurst exponent and the price change distributions

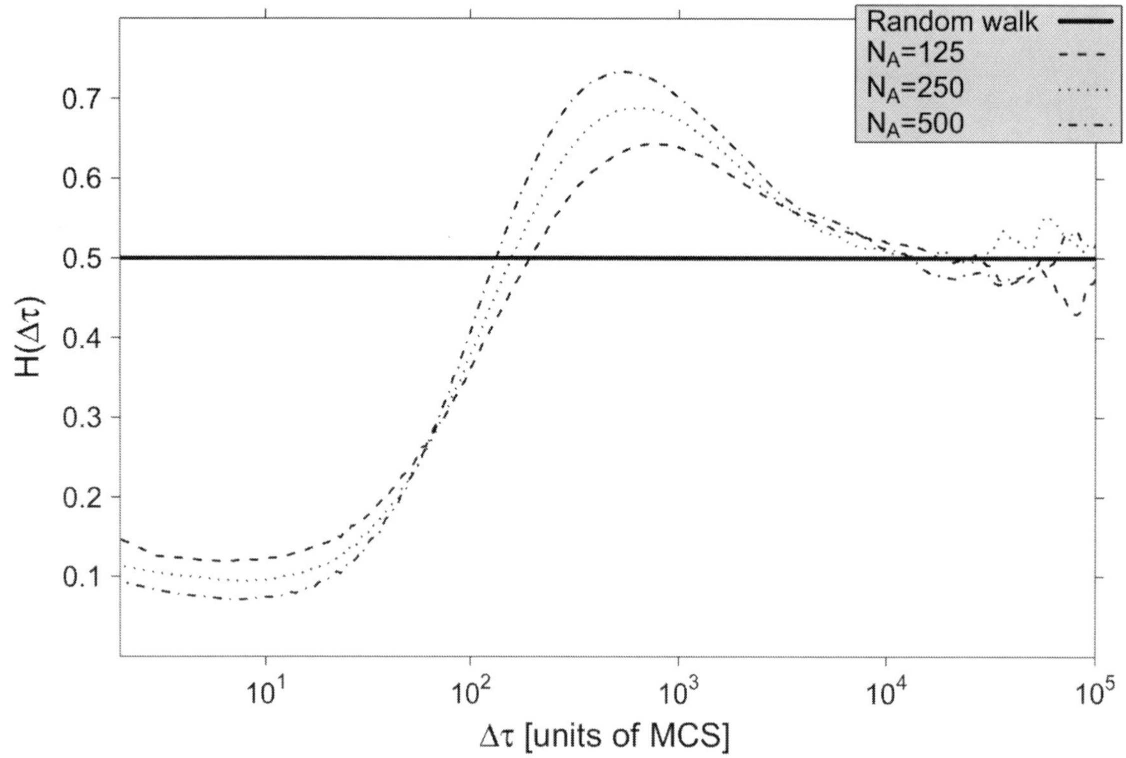

Figure 11. Dynamic order placement depth: Hurst exponent $H(\Delta\tau)$ for various values of N_A with $\lambda_0 = 100$, $C_\lambda = 10$, $\alpha = 0.15$, $\mu = 0.025$, $\delta = 0.025$, and $\Delta S = 0.001$.

are shown for coupling the order placement depth λ to the prevailing trend. For the calculation, an average is taken of 50 independent simulation runs. Each simulation run lasts 10^6 MCS. With the dynamic order placement depth, the Order Book Model can reproduce a persistent scaling behavior on medium time scales caused by the FRW mechanism and is also able to exhibit fat-tailed price change distributions. The origin of fat-tailed price change distributions is a widely discussed problem in econophysics. Often a truncated Lévy distribution [16, 17, 47] is suggested in order to approximate fat-tailed price change distributions, which can be found at real financial markets indeed. A Lévy stable distribution has the following properties: On the one hand, a Lévy stable distribution is scale invariant, on the other hand, it exhibits an infinite variance [16]. In contrast, a truncated Lévy distribution (TLD) [48, 49] exhibits a finite variance and shows scaling behavior in a large, but finite interval. As a second approach, also the possibility of power law tails in the price change distributions are widely discussed [16]. For example, it is shown in [50] for the stock market index S&P 500, that the price change distributions for time lags $\Delta\tau \leq 4$ days show a power-law behavior with an exponent $\alpha \approx 3$. And this exponent is outside the stable Lévy regime, which can be found in the range $0 < \alpha_L < 2$. For larger time lags, a slow convergence to random walk statistics is found. For this background, a TLD [49] will be used for fitting fat-tailed price changing distributions, which are obtained by the Order Book Model with dynamic order placement depth. The characteristic function of the TLD is given by

$$\Lambda_{\alpha_L,c_1,l}(f_n) = \exp\left(c_0 - c_1 \frac{(f_n^2 + 1/l^2)^{\alpha_L/2}}{\cos(\pi\alpha_L/2)} \cos\left(\alpha_L \arctan(l|f_n|)\right)\right) \quad (9)$$

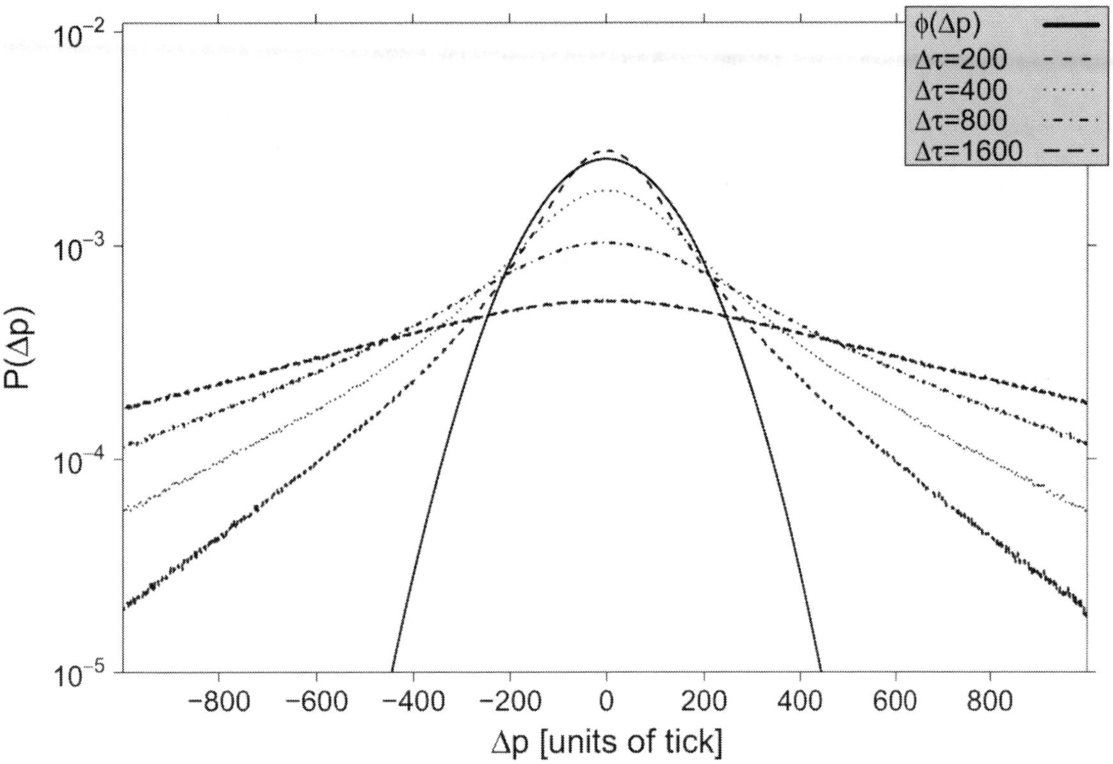

Figure 12. Dynamic order placement depth: Distributions of price changes for various values of $\Delta\tau$ with the same parameter values as in Fig. 11. A Gaussian approximation $\phi(\Delta p) = a \cdot \exp(-b \cdot \Delta p^2)$ for $\Delta\tau = 200$ with the parameters $a = 2.54 \times 10^{-3} \pm 10^{-5}$ and $b = (2.80 \pm 0.02) \times 10^{-5}$ strongly underestimates the probability for large price changes.

with scaling factors

$$c_0 = \frac{l^{-\alpha_L}}{\cos(\pi\alpha_L/2)} \tag{10}$$

and c_1. For the reason, that only the characteristic function of the TLD is given in an analytic form, a discrete Fourier transformation of the price change distributions is necessary, in order to analyze the obtained simulation data. After applying this method, one can compare the Lévy exponent α_L, which was measured for real financial data time series and which takes values in the range of ≈ 1.4–1.5 [48, 52]. For this background, the parameter selection $C_\lambda = 1$ seems to be the best approximation [40]. If one couples the order placement depth to the prevailing trend, then fat-tailed price change distributions can be observed. Furthermore, this feature is independent of the persistent behavior on medium time scales. The persistence could be obtained by the FRW and so by imposing non-stationary increments on the price process. On the other hand, it is also possible to generate fat-tailed price change distributions without $H > 1/2$ for medium time scales, such that these two properties are truly independent.

5. Summary

In the beginning of exchange based trading, floor trading was a widespread form of trading. The tulip bulb crisis in the United Provinces and the London Gold Fixing are exemplarily discussed being famous examples in the history of exchange based trading. In the course of the introduction and the progress in information technology, electronic exchanges were created.

Based on the market microstructure of modern electronic financial markets, the Order Book Model is presented, which was introduced in [39] as a multi-agent based system for modeling financial markets and which was analyzed in detail in [40]. It can be shown, that liquidity providers have a systematic advantage by the possibility of transmitting limit orders, compared to the group of the liquidity takers. This version of the Order Book Model is able to reproduce very simple facts of financial markets. With an exponentially distributed order placement depth, a log-normally distributed order book depth is obtained. The corresponding price time series possess an antipersistent price behavior on short time scales. On medium and long time scales, the Hurst exponent converges towards a random walk behavior. The price change distributions obey an almost Gaussian shape. This basic realization of the Order Book Model reflects a stationary market due to a symmetry, reached by static and identical buy and sell probabilities. However, when one additionally introduces a symmetry-breaking extension by the feedback random walk, the Order Book Model is displaced from its stationary state, in order to implement market trends on short time scales. With this perturbation, a persistent price behavior on medium time scales is obtained. Furthermore, a coupling of the characteristic order placement depth to the prevailing market trend leads to widened price change distributions. They can be fitted by a truncated Lévy distribution. With all these extensions of the Order Book Model, it is possible to demonstrate that the generation of a nontrivial Hurst exponent $H \neq 1/2$ is independent of the generation of fat-tailed price change distributions.

6. Acknowledgements

Very useful discussions with B. Brodda, S. Golke, W. Paul, M. Pfefferkorn, J. J. Schneider, H. E. Stanley, S. Trautmann, P. Virnau, and J. Zausch are gratefully acknowledged. Each of these persons supported the underlying work with expertise and encouragement. We also thank the German Research Foundation (DFG) and the Gutenberg Academy for financial support.

References

[1] T. Preis, W. Paul, J. J. Schneider: Europhysics Letters **82**, 68005 (2008)
[2] T. Preis, P. Virnau, W. Paul, J. J. Schneider: Journal of Computational Physics **228**, 4468 (2009)
[3] T. Preis, P. Virnau, W. Paul, J. J. Schneider: New Journal of Physics (2009)
[4] M. O'Hara: *Market Microstruture Theory* (Blackwell Publishing, Malden 2007)
[5] P. M. Garber: The Journal of Political Economy **97**, 535 (1997)
[6] P. M. Garber: *Famous First Bubbles: The Fundamentals of Early Manias* (MIT Press, Cambridge 2000)
[7] A. Goldgar: *Tulipmania: Money, Honor, and Knowledge in the Dutch Golden Age* (University of Chicago Press, Chicago 2007)
[8] S. Trautmann: *Investitionen. Bewertung, Auswahl und Risikomanagement* (Springer, Berlin 2006)
[9] Deutsche Börse Group: *Geschäftsbericht 2005* (Deutsche Börse Group, Frankfurt 2006)
[10] N. Kuls: Das New Yorker Börsenparkett schrumpft auf die Hälfte. In: *Frankfurter Allgemeine Zeitung* (Frankfurt, 27th September 2007) p 23
[11] M. F. M. Osborne: Operations Research **7**, 145 (1959)
[12] M. F. M. Osborne: *The Stock Market and Finance From a Physicist's Viewpoint* (Crossgar Press, Minneapolis 1995)
[13] B. Mandelbrot: The Journal of Business **36**, 394 (1963)
[14] B. Mandelbrot: The Journal of Business **37**, 393 (1964)
[15] J.-P. Bouchaud, M. Potters: *Theory of Financial Risks - From Statistical Physics to Risk Management* (Cambrigde University Press, Cambridge 2000)
[16] R. N. Mantegna, H. E. Stanley: *An Introduction to Econophysics - Correlations and Complexity in Finance* (Cambrigde University Press, Cambridge 2000)
[17] W. Paul, J. Baschnagel: *Stochastic Processes: From Physics to Finance* (Springer, Heidelberg 2000)
[18] L. Bachelier: Ph.D. thesis, Faculté des Sciences de Paris, Paris (1900)
[19] K. Itô: Proceedings of the Imperial Academy Tokyo **20**, 519 (1944)
[20] P. A. Samuelson, Industrial Management Review **6**, 13 (1965)
[21] A. Einstein: Annalen der Physik **17**, 549 (1905)
[22] F. Black, M. Scholes: Journal of Political Economy **81**, 637 (1973)

[23] R. Merton: Bell Journal of Economics and Management Science **4**, 141 (1973)
[24] F. Wang, K. Yamasaki, S. Havlin, and H. E. Stanley: Physical Review E 79, 016103 (2009)
[25] B. Podobnik, D. Horvatic, A. Petersen, and H. E. Stanley: Europhys. Lett. 85, 50003 (2009)
[26] P. Bak, M. Paczuski, M. Shubik: Physica A **246**, 430 (1997)
[27] S. Maslov: Physica A **278**, 571 (2000)
[28] S. Maslov, M. Mills: Physica A **299**, 234 (2001)
[29] D. Challet, R. Stinchcombe: Quantitative Finance **3**, 155 (2003)
[30] R. Cont, J.-P. Bouchaud: Macroeconomic Dynamics **4**, 170 (2000)
[31] D. Stauffer, T. J. P. Penna: Physica A **256**, 284 (1998)
[32] D. Stauffer, D. Sornette: Physica A **271**, 496 (1999)
[33] T. Lux, M. Marchesi: Nature **397**, 498 (1999)
[34] I. Giardina, J.-P. Bouchaud, M. Mézard: Physica A **299**, 28 (2001)
[35] A. Krawiecki, J. A. Holyst, D. Helbing: Physical Review Letters **89**, 158701 (2002)
[36] D. Challet, M. Marsili, Y.-C. Zhang: Physica A **294**, 514 (2001)
[37] M. G. Daniels, J. D. Farmer, L. Gillemot, G. Iori, E. Smith: Physical Review Letters **90**, 108102 (2003)
[38] E. Smith, J. D. Farmer, L. Gillemot, S. Krishnamurthy: Quantitative Finance **3**, 481 (2003)
[39] T. Preis, S. Golke, W. Paul, J. J. Schneider: Europhysics Letters **75**, 510 (2006)
[40] T. Preis, S. Golke, W. Paul, J. J. Schneider: Physical Review E **76**, 016108 (2007)
[41] G. A. Darbellay, D. Wuertz: Physica A **287**, 429 (2000)
[42] M. Ausloos: Physica A **285**, 48 (2000)
[43] K. E. Bassler, G. H. Gunaratne, J. L. McCauley: Physica A **369**, 343 (2006)
[44] R. Cont: Quantitative Finance **1**, 223 (2001)
[45] M. Pasquini, M. Serva: Economics Letters **65**, 275 (1999)
[46] M. Serva, U. L. Fulco, I. M. Gléria, M. L. Lyra, F. Petronia, G. M. Viswanathan: Physica A **363**, 393 (2006)
[47] J. Voit: *The Statistical Mechanics of Financial Markets* (Springer, Berlin 2003)
[48] R. N. Mantegna, H. E. Stanley: Physical Review Letters **73**, 2946 (1994)
[49] I. Koponen: Physical Review E **52**, 1197 (1995)
[50] P. Gopikrishnan, V. Plerou, L. A. N. Amaral, M. Meyer, H. E. Stanley: Physical Review E **60**, 5305 (1999)
[51] G. M. Viswanathan, U. L. Fulco, M. L. Lyra, M. Serva: Physica A **329**, 273 (2003)
[52] R. N. Mantegna, H. E. Stanley: Nature **376**, 46 (1995)

AUTHOR INDEX

Institute of Physics
Dirac House, Temple Back
Bristol BS1 6BE UK

CURRAN ASSOCIATES INC
proceedings.com